METHODS IN MOLECULAR BIOLOGY™

Series Editor
John M. Walker
School of Life Sciences
University of Hertfordshire
Hatfield, Hertfordshire, AL10 9AB, UK

For further volumes:
http://www.springer.com/series/7651

Helicobacter Species

Methods and Protocols

Edited by

JeanMarie Houghton

Division of Gastroenterology, Department of Medicine,
University of Massachusetts Medical School Worcester, Worcester, MA, USA

Humana Press

Editor
JeanMarie Houghton, MD, PhD
Division of Gastroenterology
Department of Medicine
University of Massachusetts Medical School Worcester
Worcester, MA, USA

ISSN 1064-3745 ISSN 1940-6029 (electronic)
ISBN 978-1-4939-5907-5 ISBN 978-1-62703-005-2 (eBook)
DOI 10.1007/978-1-62703-005-2
Springer New York Heidelberg Dordrecht London

Printed on acid-free paper

Humana Press is a brand of Springer
Springer is part of Springer Science+Business Media (www.springer.com)

Preface

Identifying Helicobacter infection as the leading cause of peptic ulcer disease and gastric cancer has dramatically altered our treatment of these disease states. Over the last several decades, we have come to understand that the interplay between the bacteria, the host, and the environment all contribute to the clinical outcome of infection. Research on bacterial virulence factors and mechanisms of interaction between the bacteria and host epithelial and immune cells has provided important details used to design successful therapy and to guide vaccine development efforts. In vivo studies employing various animal models have been crucial in defining the host immune response to infection and defining the events leading up to and culminating in the formation of gastric cancer. The protocols and methods used in Helicobacter research have evolved over time, and standards across the field have been established which are essential for optimal outcomes and to allow comparison of data across different laboratories. This book presents common protocols and methods for the most often used experimental approaches in Helicobacter research. While these methods and protocols represent the common practice in several Helicobacter laboratories, it must be remembered that there are often several different "correct" ways to do something. By no means are we suggesting that alternate approaches are incorrect; however, we provide an approved, accepted, and reproducible set of methods for those who currently do not have protocols in place or who wish to change protocols.

Worcester, MA, USA *JeanMarie Houghton*

Contents

Contributors

ANNA M. CERNY • *Division of Infectious Diseases and Immunology, Department of Medicine, University of Massachusetts Medical School, Worcester, MA, USA*
TIMOTHY L. COVER • *Department of Medicine and Department of Pathology, Microbiology and Immunology, Vanderbilt University School of Medicine, Nashville, TN, USA Department of Microbiology and Immunology, Vanderbilt University School of Medicine, Nashville, TN, USA; Veterans Affairs Tennessee Valley Healthcare System, Nashville, TN, USA*
MICHAEL CRAIG • *Department of Molecular and Cellular Physiology, University of Cincinnati College of Medicine, Cincinnati, OH, USA*
KATHRYN A. EATON • *Unit for Laboratory Animal Medicine, Department of Microbiology and Immunology University of Michigan Medical School Ann Arbor, MI, USA*
RICHARD L. FERRERO • *Monash Institute of Medical Research, Monash University, Melbourne, VIC, Australia*
CEU FIGUEIREDO • *IPATIMUP, Institute of Molecular Pathology and Immunology of the University of Porto, Porto, PortugalC. Figueiredo FMUP, Faculty of Medicine of the University of Porto, Porto, Portugal*
JAMES G. FOX • *Division of Comparative Medicine, Massachusetts Institute of Technology, Cambridge, MA, USA*
YEGANEH TALEBKHAN GAROOSI • *Biotechnology Research Center, Pasteur Institute of Iran, Tehran, Iran*
ZHONGMING GE • *Division of Comparative Medicine, Massachusetts Institute of Technology, Cambridge, MA, USA*
BRIAN M. GRAY • *Unit for Laboratory Animal Medicine, Department of Microbiology and Immunology University of Michigan Medical School Ann Arbor, MI, USA*
NICOLA L. JONES • *Cell Biology Program, Research Institute, Hospital for Sick Children, Toronto, ON, Canada; Departments of Pediatrics and Physiology, University of Toronto, Toronto, ON, Canada*
JEANMARIE HOUGHTON • *Division of Gastroenterology, Department of Medicine, University of Massachusetts Medical School Worcester, Worcester, MA, USA*
SAMANEH SABERI KASHANI • *Biotechnology Research Center, Pasteur Institute of Iran, Tehran, Iran*
EVELYN A. KURT-JONES • *Division of Infectious Diseases and Immunology, Department of Medicine, University of Massachusetts Medical School, Worcester, MA, USA*
MARINA LEITE • *IPATIMUP, Institute of Molecular Pathology and Immunology of the University of Porto, Porto, Portugal*
D. SCOTT MERRELL • *Department of Microbiology and Immunology, Uniformed Services University of the Health Sciences, Bethesda, MD, USA*
MARJAN MOHAMMADI • *Biotechnology Research Center, Pasteur Institute of Iran, Tehran, Iran*

STEVEN F. MOSS • *Division of Gastroenterology, Department of Medicine, Rhode Island Hospital/Brown University, Providence, RI, USA*
JENNIFER M. NOTO • *Division of Gastroenterology, Department of Medicine, Vanderbilt University Medical Center, MRB IV 1030C MRB IV, Nashville, TN, USA*
MILAN K. PATEL • *Division of Infectious Diseases and Immunology, Department of Medicine, University of Massachusetts Medical School, Worcester, MA, USA*
RICHARD M. PEEK JR. • *Division of Gastroenterology, Department of Medicine, Vanderbilt University Medical Center, Nashville, TN, USA*
DEEPA RAJU • *Cell Biology Program, Research Institute, Hospital for Sick Children, Toronto, ON, Canada; Departments of Paediatrics and Physiology, University of Toronto, Toronto, ON, Canada*
DAVID RIZZUTI • *Cell Biology Program, Research Institute, Hospital for Sick Children, Toronto, ON, Canada*
ARLIN B. ROGERS • *Lineberger Comprehensive Cancer Center and Department of Pathology and Laboratory Medicine, University of North Carolina, Chapel Hill, NC, USA*
GLENNICE N. RYAN • *Division of Infectious Diseases and Immunology, Department of Medicine, University of Massachusetts Medical School, Worcester, MA, USA*
MICHAEL SCHUMACHER • *Department of Molecular and Cellular Physiology, University of Cincinnati College of Medicine, Cincinnati, OH, USA*
PHILIP SUTTON • *Centre for Animal Biotechnology, School of Veterinary Science, The University of Melbourne, Melbourne, VIC, Australia*
NANCY S. TAYLOR • *Division of Comparative Medicine, Massachusetts Institute of Technology, Cambridge, MA, USA*
SAHAR JAHANGIRI TAZEHKAND • *Biotechnology Research Center, Pasteur Institute of Iran, Tehran, Iran*
MELANIE I. TROMBLY • *Division of Infectious Diseases and Immunology, Department of Medicine, University of Massachusetts Medical School, Worcester, MA, USA*
MARK T. WHARY • *Division of Comparative Medicine, Massachusetts Institute of Technology, Cambridge, MA, USA*
JEANNETTE M. WHITMIRE • *Department of Microbiology and Immunology, Uniformed Services University of the Health Sciences, Bethesda, MD, USA*
JOHN E. WILSON • *School of Biotechnology and Biomolecular Sciences, The University of NSW, Kensington, NSW, Australia*
YANA ZAVROS • *Department of Molecular and Cellular Physiology, University of Cincinnati College of Medicine, Cincinnati, OH, USA*
SONGHUA ZHANG • *Division of Gastroenterology, Department of Medicine, Rhode Island Hospital/Brown University, Providence, RI, USA*

Chapter 1

Introduction

JeanMarie Houghton

Abstract

Helicobacter infection is a chronic persistent condition which is responsible for the majority of cases of gastric and duodenal ulcers, and gastric cancer. The study of the bacteria, the interaction of the bacteria with the host, and the host immune response has greatly benefited from standardization of culture techniques and animal models. The following chapters will describe the clinical aspects of infection and touch on the important techniques for optimal investigation of this infection.

Key words: Clinical disease, Helicobacter organism, Animal models

Since the original description by Warren and Marshall in 1883, of a spiral bacterium infecting the stomach of humans, Helicobacter pylori *(H. pylori)* has been recognized as a leading bacterial infection in man (1). Indeed, *H. pylori* is one of the most common chronic bacterial infections in man, (2, 3) and has documented infection dates back as far as 50,000 years with evidence of infection found in every geographic area of the world. The bacterium has adapted to growth within the human stomach to allow effective colonization yet producing an ineffective host immune response which does not allow bacterial eradication. For the most part, human and bacterium live in harmony with the majority of patients relatively asymptomatic. However, infection with *H. pylori* remains the leading cause of gastritis, peptic ulcer disease, mucosa associated lymphoid tissue (MALT) lymphoma, and gastric adenocarcinoma. Gastric adenocarcinoma is the second most common cause of cancer-related mortality worldwide and the 14th overall cause of death (4). Even though there is a very strong association between infection and disease, only approximately 10% of patients will develop peptic ulcer disease, and less than 1–2% of patients will develop gastric adenocarcinoma as a result of infection (though the cancer risk varies widely by geographical location) (5, 6). As a leading risk factor for both ulcer disease and gastric cancer, infection remains a

JeanMarie Houghton (ed.), *Helicobacter Species: Methods and Protocols*, Methods in Molecular Biology, vol. 921, DOI 10.1007/978-1-62703-005-2_1, © Springer Science+Business Media, LLC 2012

potential point of intervention for prevention and treatment of disease.

Because of the enormous health impact of infection, there is a critical need for investigation which centers on understanding the virulence of the bacterium, the dynamics of bacterial–host interaction and the host immune response to the bacterium. In order to properly accomplish these goals, a number of laboratories have developed model systems and perfected experimental techniques which allow uniformity of study design and comparison of results between groups. Below, the history and relevance of Hp infection is briefly described in order to properly place infection into the context of human disease and to succinctly detail some experimental systems which are necessary for meaningful investigation.

1. Overview of Clinical Disease

H. pylori is a micro-aerophilic spiral-shaped Gram-negative bacterium that colonizes the stomach for almost the entire lifetime of the host. *H. pylori* has been classified by the World Health Organization (WHO) as a class 1 carcinogen, although the precise mechanism by which this bacterium causes gastric cancer is not clear (7). The present thinking is that cancer arises through the combination of infection with a virulent organism and a susceptible host. Direct *H. pylori* bacterial factors, the components of the host immune response which are influenced by host genetics and altered by other infectious agents or environmental components, dietary cofactors, and hormones all interact to influence outcomes of infection. The interaction of complex components stress the need for controlled experimental systems to tease out effects of individual components of the infection environment as well as in vivo systems to test the interplay between factors.

2. The Bacteria

After the discovery that *H. pylori* was the leading cause of gastric and duodenal ulcers, and the primary risk factor for gastric adenocarcinoma and MALT lymphoma, it became apparent that understanding the properties of the bacterium which allowed colonization and initiated the host immune response would be vital for designing and implementing strategies for combating disease. Several characteristics of the bacterium are important for adherence and colonization, proliferation of the bacterium as well as inducing proliferation of the gastric mucosal cells and ultimately evoking a harmful immune response by the host. The *cag* (cytotoxin-associated gene) pathogenicity island (*cag* PAI) is a 40 kilobase segment of DNA which contains 31 genes. Many of these genes encode components of a type 4 bacterial secretion system (8–10). The secretion system acts to deliver bacterial products into the host cell via a mechanism that resembles a syringe. Products delivered into the epithelial cell include the cag gene product CagA and peptidoglycans (11, 12). Not all strains of H. pylori express the *cag* PAI. Approximately 60% of Western isolates and almost all of the East Asian strains of *H. pylori* carry the *cag* PAI (13). CagA is commonly used as a marker for the entire *cag* locus in epidemiological studies and is a 121–145 kDa protein encoded by one of the genes (*cagA*) within the *cag* PAI. Patients infected with *cagA* positive strains have a higher incidence of peptic ulcer, atrophic gastritis, and gastric adenocarcinoma (14–17).

The use of animal models to study the effects of wild-type bacteria and bacterial mutants has helped clarify the role of CagA and the *cag PAI* in the pathogenesis of gastric cancer (18–21). Disruption of the *cag PAI* produces less gastric inflammation, fewer gastric ulcers, and a lower incidence of gasric cancer when compared with rodents infected with wild-type strains (15, 20, 22). Investigation into additional bacterial components and determining the interplay between the bacterium and the host further underscore the need of uniformity of experimental systems and the availability of detailed protocols to recreate and elaborate on present studies.

In addition to altering the interaction between the bacterium and the host, bacterial eradication would be a central goal for disease treatment and prevention. Wide usage of antibiotics has created problems with antibiotic resistance and metronidazol resistance is estimated to be near 60–70% in areas of high antibiotic use (23, 24) while resistance to macrolides, such as claritromycin is also rising (24). These findings stress the importance for research efforts to identify new chemical compounds with bactericidal or bacteriostatic properties. Therefore, optimal culture conditions for in vitro studies of bacteria, techniques for the creation and manipulation of mutant bacterial strains as well as standards for preparing bacteria for in vivo analysis are essential for advancing the field of bacterial–host interaction.

3. Animal Models

The germ-free pig was the first animal successfully colonized with Helicobacter species (25, 26). While use of the pig model was time-consuming, cumbersome, and expensive, it paved the way for the establishment of additional animal models for Helicobacter infection in order to establish the natural effects of infection of animals and later establish experimental models for research purposes. Commonly used models include the ferret (27–29), the gerbil (21), and the more frequently used mouse models of infection (8–10, 30–32). Indeed, the mouse has been instrumental in determining the effects of the host immune response to infection, analyzing the contributions of individual cytokines and chemokines to disease states (33–40) and interrogating the role of bone marrow cells to the initiation and progression of gastric cancer (41). The role of bacterial eradication in prevention and treatment of disease (42) has been shown in a mouse model and supports human studies suggesting that bacterial eradication impacts disease in humans.

In the following chapters, techniques to grow, manipulate, and analyze Helicobacter bacteria as well as methods for infection, manipulation, and analysis of animal models are presented in order to facilitate study, standardize approaches and allow comparison of results between laboratories.

References

1. Marshall BJ, Warren JR (1983) Unidentified curved bacillus on gastric epithelium in active chronic gastritis. Lancet 1:1273–5
2. Peek RM Jr, Blaser MJ (2002) Helicobacter pylori and gastrointestinal tract adenocarcinomas. Nat Rev Cancer 2:28–37
3. Kavermann H, Burns BP, Angermuller K et al (2003) Identification and characterization of helicobacter pylori genes essential for gastric colonization. J Exp Med 197:813–822
4. Parkin DM, Laara E, Muir CS (1988) Estimates of the worldwide frequency of sixteen major cancers in 1980. Int J Cancer 41:184–197
5. Parsonnet J, Friedman GD, Vandersteen DP et al (1991) Helicobacter pylori infection and the risk of gastric carcinoma. N Engl J Med 325:1127–1131
6. Blaser MJ (1992) Helicobacter pylori: its role in disease. Clin Infect Dis 15:386–391
7. IARC Working Group on the Evaluation of Carcinogenic Risks to Humans (1994) Schistosomes, liver flukes and Helicobacter pylori. Lyon, 7–14 June 1994. IARC Monogr Eval Carcinog Risks Hum 61:1–241.
8. Tomb JF, White O, Kerlavage AR et al (1997) The complete genome sequence of the gastric pathogen helicobacter pylori. Nature 388: 539–547
9. Alm RA, Ling LS, Moir DT et al (1999) Genomic-sequence comparison of two unrelated isolates of the human gastric pathogen helicobacter pylori. Nature 397:176–180
10. Censini S, Lange C, Xiang Z et al (1996) Cag, a pathogenicity island of helicobacter pylori, encodes type I-specific and disease-associated virulence factors. Proc Natl Acad Sci USA 93:14648–14653
11. Covacci A, Telford JL, Del Giudice G et al (1999) Helicobacter pylori virulence and genetic geography. Science 284:1328–1333
12. Viala J, Chaput C, Boneca IG et al (2004) Nod1 responds to peptidoglycan delivered by the helicobacter pylori Cag pathogenicity island. Nat Immunol 5:1166–1174
13. Mizushima T, Sugiyama T, Komatsu Y et al (2001) Clinical relevance of the babA2 genotype of helicobacter pylori in Japanese clinical isolates. J Clin Microbiol 39:2463–2465
14. Blaser MJ, Perez-Perez GI, Kleanthous H et al (1995) Infection with helicobacter pylori strains possessing cagA is associated with an increased risk of developing adenocarcinoma of the stomach. Cancer Res 55:2111–2115
15. Gerhard M, Lehn N, Neumayer N et al (1999) Clinical relevance of the helicobacter pylori gene for blood-group antigen-binding adhesin. Proc Natl Acad Sci USA 96:12778–12783
16. Parsonnet J, Friedman GD, Orentreich N et al (1997) Risk for gastric cancer in people with CagA positive or CagA negative helicobacter pylori infection. Gut 40:297–301
17. Hamlet A, Thoreson AC, Nilsson O et al (1999) Duodenal helicobacter pylori infection differs in cagA genotype between asymptomatic

subjects and patients with duodenal ulcers. Gastroenterology 116:259–268
18. Franco AT, Johnston E, Krishna U et al (2008) Regulation of gastric carcinogenesis by helicobacter pylori virulence factors. Cancer Res 68:379–387
19. Murata-Kamiya N, Kurashima Y, Teishikata Y et al (2007) Helicobacter pylori CagA interacts with E-cadherin and deregulates the beta-catenin signal that promotes intestinal transdifferentiation in gastric epithelial cells. Oncogene 26:4617–4626
20. Franco AT, Israel DA, Washington MK et al (2005) Activation of beta-catenin by carcinogenic helicobacter pylori. Proc Natl Acad Sci USA 102:10646–10651
21. Ogura K, Maeda S, Nakao M et al (2000) Virulence factors of helicobacter pylori responsible for gastric diseases in Mongolian gerbil. J Exp Med 192:1601–1610
22. Israel DA, Salama N, Arnold CN et al (2001) Helicobacter pylori strain-specific differences in genetic content, identified by microarray, influence host inflammatory responses. J Clin Invest 107:611–620
23. Ling TK, Cheng AF, Sung JJ et al (1996) An increase in helicobacter pylori strains resistant to metronidazole: a five-year study. Helicobacter 1:57–61
24. Iovene MR, Romano M, Pilloni AP et al (1999) Prevalence of antimicrobial resistance in eighty clinical isolates of helicobacter pylori. Chemotherapy 45:8–14
25. Krakowka S, Morgan DR, Kraft WG et al (1987) Establishment of gastric campylobacter pylori infection in the neonatal gnotobiotic piglet. Infect Immun 55:2789–2796
26. Lambert JR, Borromeo M, Pinkard KJ et al (1987) Colonization of gnotobiotic piglets with campylobacter pyloridis—an animal model? J Infect Dis 155:1344
27. Fox JG, Edrise BM, Cabot EB et al (1986) Campylobacter-like organisms isolated from gastric mucosa of ferrets. Am J Vet Res 47: 236–239
28. Fox JG, Cabot EB, Taylor NS et al (1988) Gastric colonization by campylobacter pylori subsp. Mustelae in ferrets. Infect Immun 56: 2994–2996
29. Fox JG, Otto G, Murphy JC et al (1991) Gastric colonization of the ferret with helicobacter species: natural and experimental infections. Rev Infect Dis 13(Suppl 8):S671–80
30. Karita M, Kouchiyama T, Okita K et al (1991) New small animal model for human gastric helicobacter pylori infection: success in both nude and euthymic mice. Am J Gastroenterol 86:1596–1603
31. Karita M, Li Q, Cantero D et al (1994) Establishment of a small animal model for human helicobacter pylori infection using germ-free mouse. Am J Gastroenterol 89: 208–213
32. Marchetti M, Arico B, Burroni D et al (1995) Development of a mouse model of helicobacter pylori infection that mimics human disease. Science 267:1655–1658
33. El-Omar EM, Rabkin CS, Gammon MD et al (2003) Increased risk of noncardia gastric cancer associated with proinflammatory cytokine gene polymorphisms. Gastroenterology 124: 1193–1201
34. Suzuki T, Ina K, Nishiwaki T et al (2004) Differential roles of interleukin-1beta and interleukin-8 in neutrophil transendothelial migration in patients with helicobacter pylori infection. Scand J Gastroenterol 39: 313–321
35. Bamford KB, Fan X, Crowe SE et al (1998) Lymphocytes in the human gastric mucosa during helicobacter pylori have a T helper cell 1 phenotype. Gastroenterology 114:482–492
36. Mohammadi M, Nedrud J, Redline R et al (1997) Murine CD4 T-cell response to helicobacter infection: TH1 cells enhance gastritis and TH2 cells reduce bacterial load. Gastroenterology 113:1848–1857
37. Mohammadi M, Czinn S, Redline R et al (1996) Helicobacter-specific cell-mediated immune responses display a predominant Th1 phenotype and promote a delayed-type hypersensitivity response in the stomachs of mice. J Immunol 156:4729–4738
38. Cui G, Houghton J, Finkel N et al (2003) IFN-gamma infusion induces gastric atrophy, metaplasia and dysplasia in the absence of helicobacter infection-a role for immune response in helicobacter disease. Gastroenterol. 124:4.
39. Fox JG, Beck P, Dangler CA et al (2000) Concurrent enteric helminth infection modulates inflammation and gastric immune responses and reduces helicobacter-induced gastric atrophy. Nat Med 6:536–542
40. Stoicov C, Whary M, Rogers AB et al (2004) Coinfection modulates inflammatory responses and clinical outcome of helicobacter felis and toxoplasma gondii infections. J Immunol 173:3329–3336
41. Houghton J, Stoicov C, Nomura S et al (2004) Gastric cancer originating from bone marrow-derived cells. Science 306:1568–1571
42. Cai X, Carlson J, Stoicov C et al (2005) Helicobacter felis eradication restores normal architecture and inhibits gastric cancer progression in C57BL/6 mice. Gastroenterology 128:1937–1952

Chapter 2

Helicobacter pylori: An Overview

Jennifer M. Noto and Richard M. Peek Jr.

Abstract

Infection with *Helicobacter pylori* is directly responsible for substantial morbidity and mortality worldwide. This ubiquitous organism causes disease through the interaction of multiple factors including bacterial factors, host immune responses, and environmental factors. The following chapters address the bacterial specific contributions to disease.

Key words: *H. pylori*, Gastric cancer, Histology

1. *Helicobacter pylori*

Helicobacter pylori is a Gram-negative bacterium that selectively colonizes the gastric epithelium of humans. In 1983, Marshall and Warren first identified *H. pylori* juxtaposed to the gastric epithelium of patients with chronic gastritis (1). They were subsequently awarded the Nobel Prize of Medicine in 2005 for their discovery of this pathogenic bacterium and for its role in peptic ulcer disease. Since that discovery, a strong link has been established between *H. pylori* and a diverse spectrum of gastrointestinal diseases, including gastric and duodenal ulceration, gastric adenocarcinoma, mucosa-associated lymphoid tissue (MALT) lymphoma, and non-Hodgkin's lymphoma of the stomach. As a result, the World Health Organization classified *H. pylori* as a class I carcinogen for gastric cancer (2) and currently this organism is considered the most common etiologic agent of infection-related cancers, representing 5.5% of the global cancer burden (3, 4).

H. pylori inhabits the human stomach for decades and recent evidence now supports the tenet that *H. pylori* has coevolved with

JeanMarie Houghton (ed.), *Helicobacter Species: Methods and Protocols*, Methods in Molecular Biology, vol. 921,
DOI 10.1007/978-1-62703-005-2_2, © Springer Science+Business Media, LLC 2012

humans for tens of thousands of years, with genetic studies indicating that humans have been colonized with *H. pylori* for at least 58,000 years (5). Present in approximately half of the world's population, *H. pylori* remains the most common bacterial infection in humans. Although the majority of individuals infected with *H. pylori* remain asymptomatic throughout their life, essentially all develop chronic inflammation (6). Among infected individuals, approximately 10% develop peptic ulcer disease, 1–3% develop gastric adenocarcinoma, and less than 0.1% develop MALT (7).

2. Gastric Adenocarcinoma

Gastric adenocarcinoma is the second leading cause of cancer-related death worldwide (8–14), accounting for approximately 700,000 deaths each year (3, 4, 11). In some regions of the world, gastric adenocarcinoma is the most common malignancy, and in Japan, the incidence of gastric cancer is almost tenfold higher than rates observed in the United States. There are two histologically distinct forms of gastric adenocarcinoma, diffuse and intestinal type. Diffuse type cancer is characterized by an infiltration of gastric mucosa with neoplastic cells, while intestinal type cancer progresses through a series of well-defined pathological steps, first described in 1975 (15). Intestinal-type adenocarcinoma involves the progression from normal gastric mucosa to chronic superficial gastritis, followed by atrophic gastritis and intestinal metaplasia, and finally leading to dysplasia and adenocarcinoma (16, 17) (Fig. 1). Chronic gastric inflammation is the primary step in this cascade and is induced by *Helicobacter pylori*, which is the strongest known risk factor for the development of premalignant lesions that occur later in the carcinogenic cascade (atrophic gastritis, intestinal metaplasia, dysplasia, and gastric adenocarcinoma) (8–14). Gastritis of the corpus predisposes towards gastric cancer, which is thought to be due, in part, to decreased acid secretion or hypoacidity. In contrast, inflammation in the antrum with relative sparing of the acid-secreting corpus results in increased acid production or hyperacidity, and predisposes to duodenal ulceration, which is associated with a decreased risk of gastric cancer (18). Eradication of *H. pylori* significantly decreases the risk of gastric adenocarcinoma in patients without premalignant lesions (19), and reduces the severity of premalignant lesions (20). Collectively, these studies provide clear evidence that *Helicobacter pylori* initiates the transformation of normal gastric mucosa towards gastric adenocarcinoma. However, although more than half of the world's population is infected with *H. pylori*, only a fraction of individuals ever develop

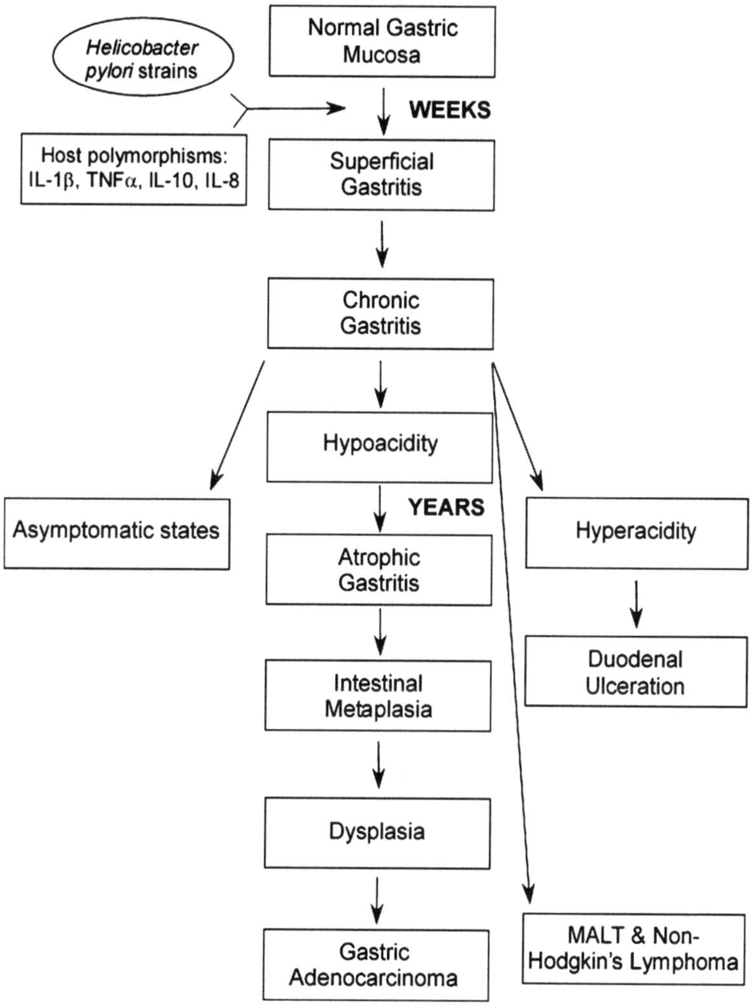

Fig. 1. Progression to intestinal-type gastric adenocarcinoma. *H. pylori* colonizes human gastric mucosa, leading to superficial acute and chronic gastritis over time. *H. pylori* strains that express the *cag* pathogenicity island and host polymorphisms that promote high expression levels of pro-inflammatory cytokines augment the risk for developing atrophic gastritis, intestinal metaplasia, dysplasia, and, ultimately, gastric adenocarcinoma.

dysplasia or gastric adenocarcinoma. The variable outcomes of *H. pylori* infection depend on strain-specific bacterial constituents, inflammatory responses governed by host genetic diversity, or environmental influences, which ultimately influence the interactions between *H. pylori* and its human host. The next several chapters will explore techniques to investigate and manipulate the organism itself as a means of interrogating bacterial specific factors involved in human and marine disease.

References

1. Marshall BJ, Warren JR (1984) Unidentified curved bacilli in the stomach of patients with gastritis and peptic ulceration. Lancet 1:1311–1315
2. International Agency for Research on Cancers (1994) Monographs on the evaluation of caci-nogenic risks to humans. World Health Organization, Geneva
3. Parkin DM (2006) The global health burden of infection-associated cancers in the year 2002. Int J Cancer 118:3030–3044
4. Parkin DM, Bray F, Ferlay J, Pisani P (2005) Global cancer statistics, 2002. CA Cancer J Clin 55:74–108
5. Linz B, Balloux F, Moodley Y, Manica A, Liu H, Roumagnac P et al (2007) An African origin for the intimate association between humans and *Helicobacter pylori*. Nature 445:915–918
6. Peek RM (2002) New insights into microbially initiated gastric malignancies: beyond the usual suspects. Gastroenterology 123:1739–1740, discussion 1740–1731
7. Peek RM, Crabtree JE (2006) *Helicobacter* infection and gastric neoplasia. J Pathol 208:233–248
8. Bechi P, Balzi M, Becciolini A, Maugeri A, Raggi CC, Amorosi A et al (1996) *Helicobacter pylori* and cell proliferation of the gastric mucosa: possible implications for gastric carcinogenesis. Am J Gastroenterol 91:271–276
9. Beswick EJ, Suarez G, Reyes VE (2006) *H. pylori* and host interactions that influence pathogenesis. World J Gastroenterol 12:5599–5605
10. Blanchard TG, Drakes ML, Czinn SJ (2004) *Helicobacter* infection: pathogenesis. Curr Opin Gastroenterol 20:10–15
11. Correa P (2004) Is gastric cancer preventable? Gut 53:1217–1219
12. Ernst PB, Peura DA, Crowe SE (2006) The translation of *Helicobacter pylori* basic research to patient care. Gastroenterology 130: 188–206
13. Moss SF, Blaser MJ (2005) Mechanisms of disease: inflammation and the origins of cancer. Nat Clin Pract Oncol 2:90–97
14. Peek RM, Blaser MJ (2002) *Helicobacter pylori* and gastrointestinal tract adenocarcinomas. Nature Rev Cancer 2:28–37
15. Correa P (1992) Human gastric carcinogenesis: a multistep and multifactorial process- First American Cancer Society Award Lecture on Cancer Epidemiology and Prevention. Cancer Res 52:6735–6740
16. Correa P (1996) *Helicobacter pylori* and gastric cancer: state of the art. Cancer Epidemiol Biomarkers Prev 5:477–481
17. Sipponen P, Marshall BJ (2000) Gastritis and gastric cancer Western countries. Gastroenterol Clin North Am 29:579–592
18. Atherton JC (2006) The pathogenesis of *Helicobacter pylori*-induced gastro-duodenal diseases. Annu Rev Pathol 1:63–96
19. Wong BC, Lam SK, Wong WM, Chen JS, Zheng TT, Feng RE et al (2004) *Helicobacter pylori* eradication to prevent gastric cancer in a high-risk region of China: a randomized controlled trial. JAMA 291:187–194
20. Mera R, Fontham ET, Bravo LE, Bravo JC, Piazuelo MB, Camargo MC et al (2005) Long term follow up of patients treated for *Helicobacter pylori* infection. Gut 54: 1536–1540

Chapter 3

Perspectives on Methodology for In Vitro Culture of *Helicobacter pylori*

Timothy L. Cover

Abstract

Over the past 25 years, a variety of methods have been developed for culture of *Helicobacter pylori* in vitro. *H. pylori* is a capnophilic and microaerophilic organism that is typically cultured using complex culture media. Analysis of *H. pylori* growth in chemically defined media has provided insight into the nutritional requirements, physiology, and metabolic capacities of this organism.

Key words: *Helicobacter pylori*, Defined medium, Capnophilic, Microaerophilic, Nutritional requirements, Nutrient acquisition

1. Discovery of Helicobacter in Gastric Specimens

Spiral-shaped bacteria were visualized in histologic sections of human gastric specimens throughout most of the twentieth century, but these organisms remained uncharacterized until 1984, when gastric bacteria (known now as *Helicobacter pylori*) were successfully cultured in vitro for the first time by Marshall and Warren (1). These investigators isolated *H. pylori* by placing minced human gastric tissue on nonselective media and culturing at 37°C under microaerobic conditions for 4 days, using methods similar to those used for isolation of Campylobacter species. Over the past 25 years, there have been various improvements in the methods for culture of *H. pylori* from human gastric specimens, but the general approach remains similar to that which was used initially in 1984. General principles include the use of a rich culture medium containing blood or serum, a microaerobic and hypercarbic atmosphere, high humidity, temperature of 37°C, and incubation periods ranging

JeanMarie Houghton (ed.), *Helicobacter Species: Methods and Protocols*, Methods in Molecular Biology, vol. 921,
DOI 10.1007/978-1-62703-005-2_3, © Springer Science+Business Media, LLC 2012

from 4 to 10 days (2). *H. pylori* can be cultured from gastric specimens using nonselective media, but antibiotics are often added to allow selective growth of *H. pylori*. For example, combinations of vancomycin, polymyxin, trimethoprim, bacitracin, nalidixic acid, and amphotericin are commonly included in *H. pylori*-selective media.

2. Growth of Bacteria for Laboratory Study on Agar Plates

After recovery of *H. pylori* from human gastric biopsy specimens, the bacteria can be propagated in vitro using a variety of approaches. Commonly used media include Brucella agar, Columbia agar, brain–heart infusion agar, or trypicase soy agar as the base, supplemented with sheep blood or horse blood (5–10%) (2). Growth of *H. pylori* on serum-free medium can be accomplished by substituting β-cyclodextrin in place of blood products (3). In addition, egg yolk emulsion medium has been described as a blood-free medium for growth of *H. pylori* (4).

H. pylori is a capnophilic organism that requires an atmosphere enriched in CO_2 (typically 5–10%) for growth (5, 6). The requirement for high CO_2 concentrations is probably related to multiple factors. For example, production of pyruvate through CO_2 fixation may provide a route for carbon assimilation (7), and CO_2 may have a role in pH homeostasis since *H. pylori* carbonic anhydrase catalyzes interconversion of CO_2 and HCO_3^- (8). *H. pylori* is an oxygen-sensitive microaerophile (5), and consequently, microaerobic conditions are used when initially culturing *H. pylori* from gastric biopsy specimens. The sensitivity of *H. pylori* to oxygen is attributed to oxygen-dependent inactivation of essential bacterial enzymes (6). When present in high cell densities, laboratory-adapted strains of *H. pylori* can grow in a range of atmospheric oxygen tensions ranging from microaerobic (<5% oxygen) to fully aerobic (21% oxygen) (5). Several characteristics of *H. pylori*, including hemolysin production, metronidazole resistance, ferredoxin oxidoreductase activity, and ability of the bacteria to induce alterations in epithelial cells, are reported to differ depending on whether the bacteria are grown in microaerobic or aerobic conditions (5, 9–11).

3. Growth of Bacteria for Laboratory Study: Liquid Culture

When growth of *H. pylori* in liquid culture is required, a commonly used medium is Brucella broth supplemented with fetal bovine serum (5–10%). As an alternative to fetal bovine serum, β-cyclodextrin, charcoal, or starch may be used (12). The exact mechanisms by which serum, β-cyclodextrin, charcoal, or starch promote growth of *H. pylori* have not been investigated in detail. One possibility is that serum may contain growth-stimulating factors. β-cyclodextrin, charcoal, and starch may bind fatty acids or other toxic metabolites produced by the bacteria (13, 14). Growth of *H. pylori* in liquid culture on a large scale has been successfully accomplished by use of a fermentor (15, 16).

For some experiments, it is desirable to culture *H. pylori* using a defined medium instead of the complex media described above. In general, defined media for culture of *H. pylori* consist of various salts, a purine, vitamins, amino acids, and trace metals (17–20). To enhance bacterial growth, the defined media can be supplemented with either bovine serum albumin or a mixture of β-cyclodextrin and cholesterol (17, 19, 20). Initial formulations of defined media for growth of *H. pylori* had a composition similar to that found in commercially available RPMI tissue culture medium (17), whereas the composition of more recent formulations is similar to that found in F12 medium (20). The use of defined media for culture of *H. pylori* has allowed important insights into the nutritional requirements, physiology, and metabolism of this organism (17–23), and also has facilitated analysis of the *H. pylori* extracellular proteome (24, 25).

Nearly all *H. pylori* strains require arginine, histidine, leucine, methionine, phenylalanine, and valine for growth (17–20). Interestingly, the amino acids that are essential for *H. pylori* growth are similar to the amino acids that are essential for humans (26). There is variability among strains in a requirement for alanine, serine, cysteine, proline, and isoleucine (17–20). Other factors required for *H. pylori* growth include pyruvate, thiamine, and hypoxanthine (20, 21), as well as several metals, including iron, zinc, and magnesium (21). These nutritional requirements correlate with the absence of the corresponding biosynthetic genes in the *H. pylori* genome (27).

4. In Vitro and In Vivo Growth Requirements of *H. pylori* Are Different

Although current culture methods provide valuable insights into the biology of *H. pylori*, it is important to recognize the limitations of these systems in replicating conditions present in the human stomach. Numerous bacterial proteins are required for *H. pylori* growth in vivo but are nonessential in vitro (28). For example, urease (a nickel-containing enzyme) is required for *H. pylori* colonization of the stomach (29, 30) but is not required for *H. pylori* growth in vitro. Correspondingly, nickel is probably required for *H. pylori* growth in vivo, but does not seem to be required for

H. pylori growth in vitro (21). Mechanisms of nutrient acquisition also may differ during bacterial growth in vivo compared to in vitro. For example, *H. pylori* utilizes free iron when cultured in vitro (21) but utilizes hemoglobin, transferrin, and lactoferrin as iron sources in vivo (31). These examples illustrate that there are likely to be important differences in nutritional requirements, nutrient acquisition, and metabolism of *H. pylori* in vivo compared to in vitro. The following chapters will discuss the particular methods used by various laboratories to culture Hp.

Acknowledgments

Supported by National Institutes of Health (R01 AI039657, R01 AI068009, P01 CA116087) and Department of Veterans Affairs.

References

1. Marshall BJ, Warren JR (1984) Unidentified curved bacilli in the stomach of patients with gastritis and peptic ulceration. Lancet 1:1311–1315
2. Ndip RN, MacKay WG, Farthing MJ, Weaver LT (2003) Culturing *Helicobacter pylori* from clinical specimens: review of microbiologic methods. J Pediatr Gastroenterol Nutr 36:616–622
3. Olivieri R, Bugnoli M, Armellini D, Bianciardi S, Rappuoli R, Bayeli PF, Abate L, Esposito E, De Aziz GLJ et al (1993) Growth of *Helicobacter pylori* in media containing cyclodextrins. J Clin Microbiol 31:160–162
4. Westblom TU, Madan E, Midkiff BR (1991) Egg yolk emulsion agar, a new medium for the cultivation of *Helicobacter pylori*. J Clin Microbiol 29:819–821
5. Bury-Mone S, Kaakoush NO, Asencio C, Megraud F, Thibonnier M, De Reuse H, Mendz GL (2006) Is *Helicobacter pylori* a true microaerophile? Helicobacter 11:296–303
6. Kelly DJ (2001) The physiology and metabolism of and *Campylobacter jejuni* and *Helicobacter pylori*. Symp Ser Soc Appl Microbiol 30:16S–24S
7. St Maurice M, Cremades N, Croxen MA, Sisson G, Sancho J, Hoffman PS (2007) Flavodoxin:quinone reductase (FqrB): a redox partner of pyruvate:ferredoxin oxidoreductase that reversibly couples pyruvate oxidation to NADPH production in *Helicobacter pylori* and *Campylobacter jejuni*. J Bacteriol 189: 4764–4773
8. Bury-Mone S, Mendz GL, Ball GE, Thibonnier M, Stingl K, Ecobichon C, Ave P, Huerre M, Labigne A, Thiberge JM, De Reuse H (2008) Roles of alpha and beta carbonic anhydrases of *Helicobacter pylori* in the urease-dependent response to acidity and in colonization of the murine gastric mucosa. Infect Immun 76:497–509
9. Xia HX, Keane CT, O'Morain CA (1994) Culture of *Helicobacter pylori* under aerobic conditions on solid media. Eur J Clin Microbiol Infect Dis 13:406–409
10. Cederbrant G, Kahlmeter G, Ljungh A (1992) Proposed mechanism for metronidazole resistance in *Helicobacter pylori*. J Antimicrob Chemother 29:115–120
11. Cottet S, Corthesy-Theulaz I, Spertini F, Corthesy B (2002) Microaerophilic conditions permit to mimic in vitro events occurring during in vivo *Helicobacter pylori* infection and to identify Rho/Ras-associated proteins in cellular signaling. J Biol Chem 277:33978–33986
12. Buck GE, Smith JS (1987) Medium supplementation for growth of *Campylobacter pyloridis*. J Clin Microbiol 25:597–599
13. Hazell SL, Graham DY (1990) Unsaturated fatty acids and viability of *Helicobacter (Campylobacter) pylori*. J Clin Microbiol 28:1060–1061
14. Khulusi S, Ahmed HA, Patel P, Mendall MA, Northfield TC (1995) The effects of unsaturated fatty acids on *Helicobacter pylori* in vitro. J Med Microbiol 42:276–282

15. Marchini A, Massari P, Manetti R, Olivieri R (1994) Optimized conditions for the fermentation of *Helicobacter pylori* and production of vacuolating cytotoxin. FEMS Microbiol Lett 124:55–59
16. Deshpande M, Calenoff E, Daniels L (1995) Rapid large-scale growth of *Helicobacter pylori* in flasks and fermentors. Appl Environ Microbiol 61:2431–2435
17. Reynolds DJ, Penn CW (1994) Characteristics of *Helicobacter pylori* growth in a defined medium and determination of its amino acid requirements. Microbiology 140:2649–2656
18. Nedenskov P (1994) Nutritional requirements for growth of *Helicobacter pylori*. Appl Environ Microbiol 60:3450–3453
19. Albertson N, Wenngren I, Sjostrom JE (1998) Growth and survival of *Helicobacter pylori* in defined medium and susceptibility to Brij 78. J Clin Microbiol 36:1232–1235
20. Testerman TL, McGee DJ, Mobley HL (2001) *Helicobacter pylori* growth and urease detection in the chemically defined medium Ham's F-12 nutrient mixture. J Clin Microbiol 39:3842–3850
21. Testerman TL, Conn PB, Mobley HL, McGee DJ (2006) Nutritional requirements and antibiotic resistance patterns of *Helicobacter* species in chemically defined media. J Clin Microbiol 44:1650–1658
22. Weinberg MV, Maier RJ (2007) Peptide transport in *Helicobacter pylori*: roles of dpp and opp systems and evidence for additional peptide transporters. J Bacteriol 189:3392–3402
23. Doherty NC, Tobias A, Watson S, Atherton JC (2009) The effect of the human gut-signalling hormone, norepinephrine, on the growth of the gastric pathogen *Helicobacter pylori*. Helicobacter 14:223–230
24. Schraw W, McClain MS, Cover TL (1999) Kinetics and mechanisms of extracellular protein release by *Helicobacter pylori*. Infect Immun 67:5247–5252
25. Smith TG, Lim JM, Weinberg MV, Wells L, Hoover TR (2007) Direct analysis of the extracellular proteome from two strains of *Helicobacter pylori*. Proteomics 7:2240–2245
26. Schilling CH, Covert MW, Famili I, Church GM, Edwards JS, Palsson BO (2002) Genome-scale metabolic model of *Helicobacter pylori* 26695. J Bacteriol 184:4582–4593
27. Doig P, de Jonge BL, Alm RA, Brown ED, Uria-Nickelsen M, Noonan B, Mills SD, Tummino P, Carmel G, Guild BC, Moir DT, Vovis GF, Trust TJ (1999) *Helicobacter pylori* physiology predicted from genomic comparison of two strains. Microbiol Mol Biol Rev 63:675–707
28. Kavermann H, Burns BP, Angermuller K, Odenbreit S, Fischer W, Melchers K, Haas R (2003) Identification and characterization of *Helicobacter pylori* genes essential for gastric colonization. J Exp Med 197:813–822
29. Eaton KA, Brooks CL, Morgan DR, Krakowka S (1991) Essential role of urease in pathogenesis of gastritis induced by *Helicobacter pylori* in gnotobiotic piglets. Infect Immun 59:2470–2475
30. Tsuda M, Karita M, Morshed MG, Okita K, Nakazawa T (1994) A urease-negative mutant of *Helicobacter pylori* constructed by allelic exchange mutagenesis lacks the ability to colonize the nude mouse stomach. Infect Immun 62:3586–3589
31. Senkovich O, Ceaser S, McGee DJ, Testerman TL (2010) Unique host iron utilization mechanisms of *Helicobacter pylori* revealed with iron-deficient chemically defined media. Infect Immun 78:1841–1849

Chapter 4

Successful Culture Techniques for *Helicobacter* Species: General Culture Techniques for *Helicobacter pylori*

Jeannette M. Whitmire and D. Scott Merrell

Abstract

Half of the world's population is persistently infected with *Helicobacter pylori*. The chronicity of this infection ultimately elicits clinical manifestations ranging from gastritis and peptic ulcers to adenocarcinoma and MALT lymphoma. Laboratory research following the initial observations of *Helicobacter* species was greatly hindered by an inability to isolate and culture the bacteria. Thus, the ability to culture bacterial species from this genus is an extremely important step in expanding clinical knowledge and development of therapies. This chapter describes successful techniques for culturing *H. pylori* on selective horse blood agar media and in Brucella broth liquid media. Additionally, the specific growth requirements of other *Helicobacter* species are noted.

Key words: *Helicobacter*, *H. pylori*, Bacterial culture, *H. bilis*, *H. felis*, *H. hepaticus*, Microaerophilic organisms

1. Introduction

This chapter describes the techniques successfully utilized in our and other laboratories for the culturing of *H. pylori* (1–7). Due to the high rate of genetic variability of *H. pylori*, once a strain of interest has been isolated, we culture the bacterium primarily using patch and lawn growth as an alternative to streaking for single colonies. This prevents the study of single colonies that may contain unwanted mutations. Accordingly, the protocol for assembling Columbia blood agar plates enriched with 5% defibrinated horse blood and selective antibiotics is detailed along with the procedures for creating Brucella broth-based liquid cultures and preparing

JeanMarie Houghton (ed.), *Helicobacter Species: Methods and Protocols*, Methods in Molecular Biology, vol. 921,
DOI 10.1007/978-1-62703-005-2_4,

freezer stocks of *H. pylori*. Additional pointers are provided to describe the specific requirements of other species of *Helicobacter*, including *H. bilis*, *H. felis*, and *H. hepaticus*.

2. Materials (see Note 1)

2.1. Agar-Based Cultures

1. Columbia Blood Agar Base (Neogen Corporation, Lansing, MI, #7125): dissolve 21.5 g in 500 mL of H_2O and autoclave (see Note 2).
2. Defibrinated Horse Blood (Hemostat Laboratories, Dixon, CA, #DHB500).
3. 1,000× Antibiotic stock: dissolve 100 mg Trimethoprim (Sigma, St. Louis, MO, #T7883) and 160 mg Amphotericin B (Amresco Inc., Solon, OH, #0414) in 20 mL dimethyl sulfoxide (DMSO) (Sigma, #D5879). Store 1 mL aliquots at −20°C (see Note 3).
4. 200× Antibiotic Stock: dissolve 100 mg Vancomycin Hydrochloride (Amresco Inc., #0990), 50 mg Cefsulodin Sodium Salt (Sigma, #C8145), and 3.3 mg Polymyxin B Sulfate (Sigma, # P1004) in 50 mL of H_2O and sterilize with a 0.22 μm filter. Store 5 mL aliquots at −20°C (see Note 3).
5. β-Cyclodextrin (Sigma, # C4805): dissolve 1 g in 5 mL of DMSO. This solution must be made immediately before adding to the agar solution and should not be stored (see Note 4).
6. Petri Dishes (BD Falcon, Franklin Lakes, NJ, #351029).
7. Brucella Broth (Neogen Corporation, #7121): dissolve 28 g in 1 L H_2O and autoclave. Store at room temperature.

2.2. Liquid Broth-Based Cultures

1. Brucella Broth: dissolve 28 g in 1 L H_2O and autoclave. Store at room temperature.
2. Fetal Bovine Serum (FBS) (Gibco/Invitrogen, Carlsbad, CA, #10437-028).
3. Vancomycin Hydrochloride solution: dissolve 100 mg in 10 mL of H_2O and sterilize with a 0.22 μm filter. Store 1 mL aliquots at −20°C.

2.3. Freezer Stock Preparations

1. Brain Heart Infusion (BHI) (BD Diagnostic Systems, Sparks, MD, #211059): dissolve 18.5 g in 500 mL of H_2O and autoclave. Store at room temperature.
2. FBS
3. Glycerol (EMD Chemicals, Gibbstown, NJ, #4750).

3. Methods (see Note 5)

3.1. Agar-Based Cultures

3.1.1. Preparation of Agar-Based Plates

1. After autoclaving the Columbia Blood Agar, place the bottle in a 55°C water bath and allow the temperature of the agar to equilibrate (see Notes 2 and 6).
2. Remove the agar from the water bath and swirl the bottle to evenly distribute the agar.
3. Add 500μL of 1,000× Antibiotic stock to the agar (see Notes 7–10).
4. Add 2.5 mL of 200× Antibiotic stock to the agar (see Notes 7–11).
5. Add 5 mL of β-cyclodextrin solution to the agar (see Notes 8 and 12).
6. Add 25 mL of Defibrinated Horse Blood to the agar (see Notes 8 and 12–13).
7. Pour the blood agar into petri dishes such that it covers the bottom of the plate taking care to avoid bubbles (see Notes 14–15).
8. Allow the plates to cool and solidify at room temperature overnight (see Note 16).
9. Store the plates in sealed plastic bags at +4°C (see Note 17).

3.1.2. Patch Growth of H. pylori *from Freezer Stocks*

1. Remove plates from the refrigerator and allow them to reach room temperature.
2. Using a sterile pipet tip, remove a small chunk of bacteria from the freezer stock.
3. Spread the chunk onto the blood agar plate over a small quarter-sized area, allowing it to melt and briefly dry.
4. Place the plate in an Anoxomat jar (Advanced Instruments, Norwood, MA) and finger-tighten the screw to seal the jar (see Note 18).
5. Attach the jar to the Anoxomat instrument and allow it to evacuate and replace the gas in the jar to create a microaerophilic atmosphere consisting of 5% O_2, 10% CO_2, and 85% N_2 (see Note 19).
6. Place the jar in a 37°C incubator overnight (see Notes 20–21).

3.1.3. Lawn Growth of H. pylori

1. Remove patched *Helicobacter* cells grown as above using a sterile cotton swab.
2. Resuspend the cells in Brucella Broth (see Notes 22–23).
3. Take the swab and spread the resuspended cells evenly over the entire surface of a new blood agar plate.

4. Allow the plate to briefly dry, invert it, and place it in an Anoxomat jar and evacuate as described above.
5. Place the jar in a 37°C incubator overnight (see Note 21).

3.1.4. Colony Growth of H. pylori (see Note 24)

1. Remove lawned *Helicobacter* cells using a sterile cotton swab.
2. Resuspend the cells in 1 mL of Brucella Broth (see Note 23).
3. Place sterile glass beads onto a new blood agar plate (see Note 25).
4. Pipet 1–10μL of the bacterial suspension onto the plate (see Notes 26–27).
5. Shake the plate horizontally in several directions to distribute the beads and suspension across the plate.
6. Pour the beads off of the plate (see Note 28).
7. Allow the plate to briefly dry, invert it, and place it in an Anoxomat jar and evacuate as described above.
8. Place the jar in a 37°C incubator. Colonies typically form in 4–7 days.

3.2. Liquid Broth-Based Cultures

3.2.1. Preparation of Liquid Media

1. Prepare the Brucella Broth as described above and allow it to reach room temperature.
2. Place 18 mL of Brucella Broth, 2 mL of FBS, and 20 μL of the 10 mg/mL vancomycin stock solution in a 125 mL screw-top flask and swirl to mix (see Notes 29–30).

3.2.2. Standard Liquid Growth of H. pylori

1. Remove lawned *Helicobacter* cells using a sterile cotton swab.
2. Resuspend the cells in Brucella Broth (see Notes 22–23).
3. Pipet the resuspended bacterial cells into the flask containing the liquid media prepared as described above until it appears slightly cloudy.
4. Swirl the liquid culture to distribute the bacterial cells.
5. Place the liquid culture in an Anoxomat jar and evacuate as described above.
6. Place the jar in a 37°C shaking incubator set with rotations of 100 rpm overnight (see Note 31).

3.2.3. OD-Controlled Liquid Growth of H. pylori (see Note 32)

1. Remove an aliquot of the liquid culture grown overnight and obtain the OD at 600 nm (OD_{600}) (see Note 23).
2. Calculate the amount of the overnight liquid culture necessary to make an OD-controlled liquid culture at the target OD_{600}. # of mL = [(target OD_{600}) × (final culture volume)]/(OD_{600} of overnight liquid culture) (see Note 33).
3. Add the calculated volume of the overnight liquid culture to a clean, sterile flask.

4. Add enough liquid media to the flask to bring it to the final culture volume desired.
5. Swirl the liquid culture to distribute the bacterial cells.
6. Remove an aliquot and obtain the OD_{600} to verify the starting OD.
7. Place the liquid culture in an Anoxomat jar and evacuate as described above.
8. Place the jar in a 37°C shaking incubator set with rotations of 100 rpm overnight (see Note 31).

3.2.4. Quantifying Bacterial Numbers (see Fig. 1 and Note 34)

1. Remove an aliquot of the liquid culture and obtain the OD_{600}.
2. Make 7 serial ten-fold dilutions from the aliquot to obtain 10^{-1} through 10^{-7} dilutions in Brucella Broth (see Note 35).

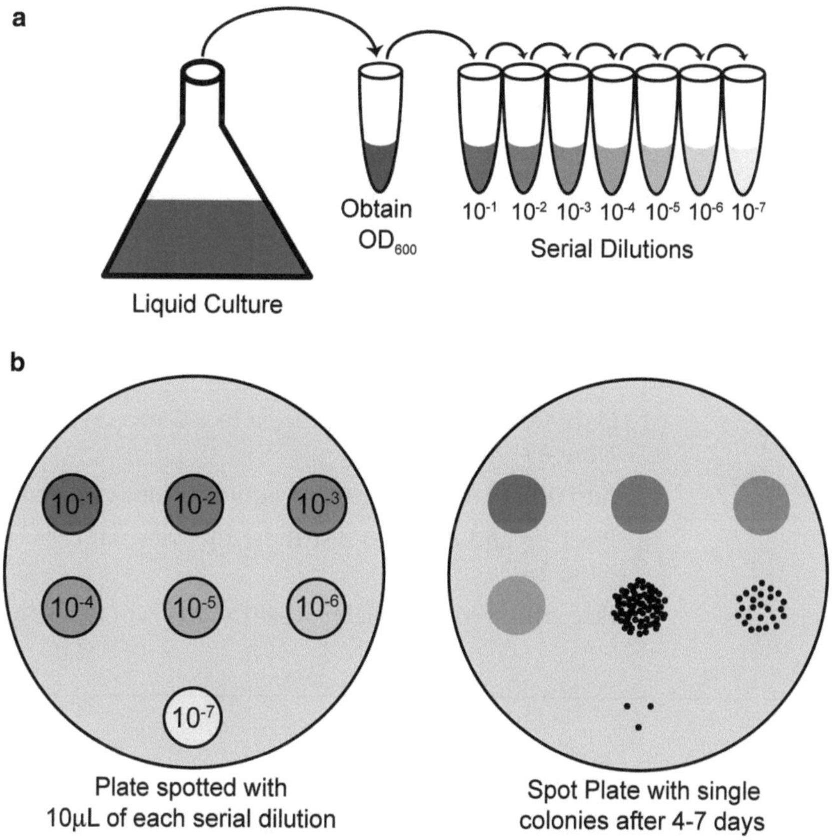

Fig. 1. Spot plating technique to quantify bacterial numbers. This figure illustrates the spot plating technique used to quantify bacterial numbers for *Helicobacter* species capable of forming single colonies as described in Subheading 3.2.4. (**a**) An aliquot is removed to determine the OD_{600}, and 7 serial ten-fold dilutions are performed in Brucella Broth to obtain 10^{-1} through 10^{-7} dilutions. (**b**) 10 μL aliquots of each dilution are then spotted onto a blood agar plate. After 4–7 days of incubation at 37°C in a microaerobic environment, single colonies are counted, and back-calculations are performed to determine the CFU/mL of the original culture (see Note 37).

3. Spot 10 μL aliquots of each dilution onto a blood agar plate (see Note 36).
4. Allow the plate to briefly dry, invert it, and place it in an Anoxomat jar and evacuate as described above.
5. Place the jar in a 37°C incubator. Colonies typically form in 4–7 days.
6. Enumerate the single colonies from each dilution that can be counted and back-calculate the CFU/mL using the following formula: [(# of colonies) × (1/dilution factor) × 1,000μL/mL]/(volume plated) (see Note 37).

3.3. Freezer Stock Preparations

3.3.1. Preparation of Freezing Media

1. Prepare the BHI as described above and allow it to reach room temperature.
2. Place 350 mL of BHI in a 1 L flask.
3. Add 50 mL of FBS to the 1 L flask.
4. Add 100 mL of glycerol to the 1 L flask.
5. Stir the solution until the glycerol is evenly distributed.
6. Sterilize the freezing media with a 0.22 μm filter.
7. Store the freezing media at +4°C.

3.3.2. Preparation of Freezer Stocks from Lawned Cells

1. Place an appropriate amount of freezing media in a 2 mL cryogenic vial (see Notes 38–39).
2. Remove lawned *Helicobacter* cells using a sterile cotton swab.
3. Resuspend the cells in the freezing media (see Note 23).
4. Place the cryogenic vial in a –80°C freezer (see Note 40).

3.3.3. Preparation of Freezer Stocks from Liquid Cultures

1. Place 900 μL of freezing media in a 2 mL cryogenic vial (see Note 39).
2. Add 900 μL of an overnight liquid culture to the vial.
3. Pipet up and down to distribute the bacterial cells evenly (see Note 23).
4. Place the cryogenic vial in a –80°C freezer (see Note 40).

4. Notes

1. All solutions requiring water should be prepared using ultrapure water with a resistance of 18.2 MΩ cm at 25°C.
2. It is helpful to make the agar solution in a 500 mL media bottle, which aids in the mixing of components and the pouring of the solution into petri dishes.
3. The combination of the 1,000× and 200× antibiotic stocks contain a mixture of the common antibiotics found in

Skirrow's *Campylobacter* Selective Supplement (Vancomycin, Trimethoprim, and Polymyxin B) (Oxoid, Cambridge, UK, # SR0069) and Dent's *Helicobacter pylori* Selective Supplement (Vancomycin, Trimethoprim, Cefsulodin, and Amphotericin B) (Oxoid, #SR0147). The final concentrations of the antibiotics within the agar are the same as the concentrations when utilizing the supplements with the exception of Amphotericin B, which is at a slightly higher concentration in the preparation presented here.

4. β-Cyclodextrin does not readily go into solution; thus, the DMSO must be added to the β-cyclodextrin and immediately and vigorously vortexed until dissolved.
5. Sterile technique must be employed for all procedures to prevent the contamination of the cultures.
6. The agar solution must be cooled to prevent the lysis of red blood cells upon addition of the horse blood to the agar.
7. Remove the aliquots of the antibiotic stocks from the −20°C freezer and allow them to reach room temperature prior to adding them to the agar solution. Cefsulodin often settles out of the solution, so vortex vigorously prior to addition.
8. After the addition of each supplement to the agar solution, swirl the bottle to evenly distribute the supplement and prevent premature solidification of the agar.
9. Final concentrations of the antibiotics are as follows: 5 μg/mL Trimethoprim, 8 μg/mL Amphotericin B, 10 μg/mL Vancomycin Hydrochloride, 5 μg/mL Cefsulodin Sodium Salt, and 0.33 μg/mL Polymyxin B Sulfate.
10. *H. bilis* and *H. hepaticus* are typically grown on commercially available Tryptic Soy agar plates supplemented with 5% sheep blood (Remel Products, Lenexa, KS, #R01200) or Brucella agar plates supplemented with 5% sheep blood and TVP (trimethoprim, vancomycin, and polymyxin B) (Remel Products, #R01258) (8–15). Alternatively, these two species have been cultured on blood agar base (Remel, #CM0055B) supplemented with 5% horse blood, 50 μg/mL Amphotericin B, 100 μg/mL Vancomycin Hydrochloride, 3.3 μg/mL Polymyxin B, 10.7 μg/mL Nalidixic Acid, and 200 μg/mL Bacitracin (15).
11. *H. felis* is not typically grown in the presence of cefsulodin (16, 17).
12. β-Cyclodextrin is at a final concentration of 0.2% and the horse blood is 5%.
13. Remove the Defibrinated Horse Blood from the refrigerator and allow it to reach room temperature prior to its addition to the agar solution. This step is important to prevent cool blood from causing the agar to solidify in the bottle.

14. Typically, one 500 mL bottle will provide enough agar for 1 sleeve containing 20–100 mm petri dishes.
15. To further reduce the number of bubbles in the agar, briefly pass the flame from a Bunsen burner over the surface of the freshly poured agar prior to solidification.
16. To help the plates dry completely after solidification, flip them over so that the lid is down and allow them to sit another 24 h.
17. The plates can typically be used up to 3 weeks after pouring, if they are stored at +4°C. However, it is important to check for contaminants, since small white colonies tend to form on the agar surface after a month in storage.
18. Alternatively, CampyGen paper sachets (Oxoid, # CN0025) can be used to generate the microaerobic atmosphere in place of the Anoxomat system.
19. *H. felis*, *H. bilis*, and *H. hepaticus* grow under microaerobic conditions comprised of 90% N_2, 5% H_2, and 5% CO_2 (10, 12, 18–20) as well as 80% N_2, 10% H_2, and 10% CO_2 (9, 14, 15, 21). These proportions of gases leave a trace amount of O_2, thereby creating the microaerobic atmosphere necessary for adequate growth of the bacteria.
20. *H. felis* (17) as well as some strains of *H. pylori* require 48 h of growth on a patch.
21. Do not allow the *Helicobacter* cells to grown on plates for an extensive period of time, since the cells will display coccoid morphology and will not grow well in subsequent cultures.
22. The amount of Brucella Broth to be used varies from 200 μL to 1 mL based on the density of the growth in the patch.
23. Prior to continuing the culture of *H. pylori* or preparing freezer stocks, it is vital to inspect the morphology of the bacterial cells using a microscope to identify any contaminants. Simply place a 5 μL aliquot on a clean, glass slide, add a cover slip, and view at 400× magnification using phase contrast. Additionally, this step is important to detect the presence of coccoid *Helicobacter* cells, which increase in numbers as cultures age and can affect downstream cultures if a large proportion of the population is in this growth stage.
24. *H. felis* and *H. hepaticus* do not grow as single colonies, but rather form a film across the inoculated plate (8, 16, 17, 22).
25. It is useful to combine equal volumes of 2 sizes of glass beads, 3 mm and 4 mm (Walter Stern, Inc., Port Washington, NY, #100C & #100E), in a 250 mL media bottle and autoclave to sterilize.
26. If plating 5 μL or less, it is useful to place 10 μL of Brucella Broth onto the plate immediately prior to adding the bacterial

suspension. The additional fluid permits a more thorough distribution of the bacterial cells across the plate.

27. If too many cells are plated, it will be difficult to distinguish single colonies. Thus, serial dilutions may be plated as described here or by spot plating as described in Subheading 3.2.4.
28. The beads can be reused by briefly placing them in a 70% Ethanol solution, washing them thoroughly with deionized water, returning them to a 250 mL media bottle, and autoclaving.
29. Alternatively, place 4.5 mL of Brucella Broth, 500 μL of FBS, and 5 μL of the 10 mg/mL vancomycin stock solution in a 25 mL flask.
30. *Helicobacter* species can also be grown in a liquid culture comprised of BHI supplemented with 10% (v/v) Fetal Bovine Serum (23). *H. pylori*, *H. bilis*, and *H. hepaticus* are also reported to grow in Brucella Broth supplemented with only 5% Fetal Bovine Serum (18, 19, 24). Additionally, a defined liquid medium for *Helicobacter* has been developed that includes only the nutrients essential for growth (25).
31. If using the smaller 25 mL flasks and lower culture volumes, it is useful to shake the culture at 110 rpm to ensure adequate aeration.
32. To improve the reproducibility of experimental results, it is imperative to prepare OD-controlled liquid cultures from starter liquid cultures. This procedure helps to ensure that each culture contains bacterial cells in similar phases of growth and with comparable morphology.
33. For example, we typically aim for a starting OD_{600} of 0.05. Thus, if the overnight culture had an OD_{600} of 0.5, add 2 mL of the liquid culture to 18 mL of liquid media [(0.05 × 20 mL)/0.5 = 2 mL].
34. Since *H. felis* and *H. hepaticus* do not form single colonies for counting, alternative methods for measuring growth must be utilized. Blanchard et al. suggest diluting the culture of interest and using a hemacytometer to count the number of cells (16). Alternatively, Senkovich et al. and Testerman et al. have successfully used a bioluminescent-ATP assay to quantify the growth of *Helicobacter* species (25, 26).
35. These dilutions can be performed in a microtiter plate or in 1.5 mL microcentrifuge tubes.
36. Alternatively, 2 μL of the dilutions can be plated. Either way, care must be taken to effectively order the dilutions on the plate in such a way that accurate calculations can ultimately be performed (see Fig. 1).

37. For example, if 25 colonies grew for the 10^{-6} dilution plated with 10 μL, the CFU/mL would equal 2.5×10^9. $[(25 \times (1/10^{-6}) \times 1{,}000)/10] = 2.5 \times 10^9$ CFU/mL.
38. The amount of freezing media to be used varies from 1 to 2 mL based on the density of lawn growth.
39. It is important to use tubes specifically designed for −80°C freezing, such as Nunc cryotube vials (Rochester, NY, #377267).
40. *H. pylori* can be challenging to revive from old freezer stocks, so it is best to maintain freezer stocks for no longer than 2–3 years.

Acknowledgments

We thank Dr. Beth Carpenter for critical reading of the manuscript. Research in the laboratory of D. Scott Merrell is supported by R01AI065529 from NIH, R073PW from USUHS, and 300411-7.20-60393 from USMCI. The contents of this work are solely the responsibility of the authors and do not necessarily represent the official views of the NIH or Department of Defense (DoD).

References

1. Baldwin DN, Shepherd B, Kraemer P et al (2007) Identification of *Helicobacter pylori* genes that contribute to stomach colonization. Infect Immun 75:1005–1016
2. Carpenter BM, McDaniel TK, Whitmire JM et al (2007) Expanding the *Helicobacter pylori* genetic toolbox: modification of an endogenous plasmid for use as a transcriptional reporter and complementation vector. Appl Environ Microbiol 73:7506–7514
3. Gancz H, Censini S, Merrell DS (2006) Iron and pH homeostasis intersect at the level of Fur regulation in the gastric pathogen *Helicobacter pylori*. Infect Immun 74: 602–614
4. Miles S, Piazuelo MB, Semino-Mora C et al (2010) Detailed in vivo analysis of the role of *Helicobacter pylori* Fur in colonization and disease. Infect Immun 78:3073–3082
5. Mouery K, Rader BA, Gaynor EC et al (2006) The stringent response is required for *Helicobacter pylori* survival of stationary phase, exposure to acid, and aerobic shock. J Bacteriol 188:5494–5500
6. Ottemann KM, Lowenthal AC (2002) *Helicobacter pylori* uses motility for initial colonization and to attain robust infection. Infect Immun 70:1984–1990
7. Salama N, Guillemin K, McDaniel TK et al (2000) A whole-genome microarray reveals genetic diversity among *Helicobacter pylori* strains. Proc Natl Acad Sci USA 97: 14668–14673
8. Fox JG, Dewhirst FE, Tully JG et al (1994) *Helicobacter hepaticus* sp. nov., a microaerophilic bacterium isolated from livers and intestinal mucosal scrapings from mice. J Clin Microbiol 32:1238–1245
9. Fox JG, Rogers AB, Whary MT et al (2004) *Helicobacter bilis*-associated hepatitis in outbred mice. Comp Med 54:571–577
10. Fox JG, Yan LL, Dewhirst FE et al (1995) *Helicobacter bilis* sp. nov., a novel *Helicobacter* species isolated from bile, livers, and intestines of aged, inbred mice. J Clin Microbiol 33:445–454
11. Hodzic E, McKisic M, Feng S et al (2001) Evaluation of diagnostic methods for *Helicobacter bilis* infection in laboratory mice. Comp Med 51:406–412
12. Mahler M, Bedigian HG, Burgett BL et al (1998) Comparison of four diagnostic methods

for detection of *Helicobacter* species in laboratory mice. Lab Anim Sci 48:85–91

13. Shames B, Fox JG, Dewhirst F et al (1995) Identification of widespread *Helicobacter hepaticus* infection in feces in commercial mouse colonies by culture and PCR assay. J Clin Microbiol 33:2968–2972
14. Shomer NH, Dangler CA, Marini RP et al (1998) *Helicobacter bilis/Helicobacter rodentium* co-infection associated with diarrhea in a colony of *scid* mice. Lab Anim Sci 48:455–459
15. Taylor NS, Xu S, Nambiar P et al (2007) Enterohepatic *Helicobacter* species are prevalent in mice from commercial and academic institutions in Asia, Europe, and North America. J Clin Microbiol 45:2166–2172
16. Blanchard TG, Nedrud JG (2006) Laboratory maintenance of *Helicobacter* species. In: Coico R, Kowalik T, Quarles JM et al (eds) Current protocols in microbiology. Wiley-Blackwell, Hoboken
17. Danon SJ, Lee A (2001) Other gastric helicobacters and spiral organisms. In: Mobley HLT, Mendz GL, Hazell SL (eds) *Helicobacter pylori*: physiology and genetics. ASM Press, Washington, DC
18. Maggio-Price L, Bielefeldt-Ohmann H, Treuting P et al (2005) Dual infection with *Helicobacter bilis* and *Helicobacter hepaticus* in p-glycoprotein-deficient *mdr1a*$^{-/-}$ mice results in colitis that progresses to dysplasia. Am J Pathol 166:1793–1806
19. Maggio-Price L, Shows D, Waggie K et al (2002) *Helicobacter bilis* infection accelerates and *H. hepaticus* infection delays the development of colitis in multiple drug resistance-deficient (*mdr1a*$^{-/-}$) mice. Am J Pathol 160:739–751
20. Paster BJ, Lee A, Fox JG et al (1991) Phylogeny of *Helicobacter felis* sp. nov., *Helicobacter mustelae*, and related bacteria. Int J Syst Bacteriol 41:31–38
21. Ihrig M, Whary MT, Dangler CA et al (2005) Gastric *Helicobacter* infection induces a Th2 phenotype but does not elevate serum cholesterol in mice lacking inducible nitric oxide synthase. Infect Immun 73:1664–1670
22. Schauer DB (2001) Enterohepatic *Helicobacter* species. In: Mobley HLT, Mendz GL, Hazell SL (eds) *Helicobacter pylori*: physiology and genetics. ASM Press, Washington, DC
23. Dhaenens L, Szczebara F, Van Nieuwenhuyse S et al (1999) Comparison of iron uptake in different *Helicobacter* species. Res Microbiol 150:475–481
24. Douraghi M, Kashani SS, Zeraati H et al (2010) Comparative evaluation of three supplements for *Helicobacter pylori* growth in liquid culture. Curr Microbiol 60:254–262
25. Testerman TL, Conn PB, Mobley HL et al (2006) Nutritional requirements and antibiotic resistance patterns of *Helicobacter* species in chemically defined media. J Clin Microbiol 44:1650–1658
26. Senkovich O, Ceaser S, McGee DJ et al (2010) Unique host iron utilization mechanisms of *Helicobacter pylori* revealed with iron-deficient chemically defined media. Infect Immun 78:1841–1849

Chapter 5

Successful Culture Techniques for *Helicobacter* Species: Establishing *H. pylori* Cultures from Infected Rodents

Jeannette M. Whitmire and D. Scott Merrell

Abstract

This chapter describes protocols for establishing *H. pylori* cultures from infected rodents.

Key words: *Helicobacter*, *H. pylori*, Bacterial culture

1. Introduction

The continued expansion of *Helicobacter* research led to the establishment of several animal models of infection to further investigate host and bacterial factors important for the disease process. Researchers developed animal models employing not only nonhuman primates but also the less sentient species of mice and gerbils, which provide an invaluable resource for the study of *H. pylori*-induced diseases (as reviewed in (1, 2)). Protocols are presented detailing the establishment of *H. pylori* strains from infected rodents.

2. Materials (see Note 1)

1. Columbia Blood Agar Base (Neogen Corporation, Lansing, MI, #7125): dissolve 21.5 g in 500 mL of H_2O and autoclave (see Note 2).
2. Defibrinated Horse Blood (Hemostat Laboratories, Dixon, CA, #DHB500).

JeanMarie Houghton (ed.), *Helicobacter Species: Methods and Protocols*, Methods in Molecular Biology, vol. 921, DOI 10.1007/978-1-62703-005-2_5, © Springer Science+Business Media, LLC 2012

3. 1,000× Antibiotic Stock: dissolve 100 mg Trimethoprim (Sigma, St. Louis, MO, #T7883) and 160 mg Amphotericin B (Amresco Inc., Solon, OH, #0414) in 20 mL dimethyl sulfoxide (DMSO) (Sigma, #D5879). Store 1 mL aliquots at −20 °C (see Note 3).
4. 200× Antibiotic Stock: dissolve 100 mg Vancomycin Hydrochloride (Amresco Inc., #0990), 50 mg Cefsulodin Sodium Salt (Sigma, #C8145), and 3.3 mg Polymyxin B Sulfate (Sigma, # P1004) in 50 mL of H_2O and sterilize with a 0.22 μm filter. Store 5 mL aliquots at −20 °C (see Note 3).
5. β-Cyclodextrin (Sigma, # C4805): dissolve 1 g in 5 mL of DMSO. This solution must be made immediately before adding to the agar solution and should not be stored (see Note 4).
6. Vancomycin Hydrochloride solution: dissolve 50 mg in 1 mL of water and sterilize with a 0.22 μm filter.
7. Nalidixic Acid solution (Sigma, #N8878): dissolve 5 mg in 0.5 mL 100% Ethanol.
8. Bacitracin solution (USB Corporation, Cleveland, OH #11805): dissolve 50 mg in 1 mL 100% Ethanol.
9. Petri Dishes (BD Falcon, Franklin Lakes, NJ, #351029).
10. Brucella Broth (Neogen Corporation, #7121): dissolve 28 g in 1 L H_2O and autoclave. Store at room temperature.

3. Methods (see Note 5)

3.1. Preparation of Agar-Based Plates

1. After autoclaving the Columbia Blood Agar, place the bottle in a 55 °C water bath and allow the temperature of the agar to equilibrate (see Notes 2 and 6).
2. Remove the agar from the water bath and swirl the bottle to evenly distribute the agar.
3. Add 500 μL of 1,000× Antibiotic stock to the agar (see Notes 7–9).
4. Add 2.5 mL of 200× Antibiotic stock to the agar (see Notes 7–9).
5. Add the entire volumes of the Vancomycin Hydrochloride, Nalidixic Acid, and Bacitracin solutions (see Notes 8–10).
6. Add 5 mL of β-cyclodextrin solution to the agar (see Notes 8 and 11).
7. Add 25 mL of Defibrinated Horse Blood to the agar (see Notes 8, 11, and 12).

8. Pour the blood agar into petri dishes such that it covers the bottom of the plate taking care to avoid bubbles (see Notes 13 and 14).
9. Allow the plates to cool and solidify at room temperature overnight (see Note 15).
10. Store the plates in sealed plastic bags at +4 °C (see Note 16).

3.2. Establishing H. pylori Cultures from Infected Rodents

1. Remove the stomach of an infected animal.
2. Bisect the glandular portion of the stomach along the greater curvature and remove any food debris.
3. Determine the weight of the stomach.
4. Place the stomach into a sterile tube containing 1 mL of Brucella Broth and mechanically homogenize (Tissue Tearor, Biospec Products, Inc., Bartlesville, OK) thoroughly to totally disrupt the tissue.
5. Place sterile glass beads onto a new blood agar plate (see Note 17).
6. Pipet 1, 10, and 100 μL of the homogenate onto separate plates (see Notes 18 and 19).
7. Shake the plates horizontally in several directions to distribute the beads and suspension across the plate.
8. Pour the beads off of the plates (see Note 20).
9. Allow the plates to dry, invert them, and place them in an Anoxomat jar (Advanced Instruments, Norwood, MA) and finger-tighten the screw to seal the jar (see Note 21).
10. Attach the jar to the Anoxomat instrument and allow it to evacuate and replace the gas in the jar to create a microaerophilic atmosphere consisting of 5% O_2, 10% CO_2, and 85% N_2.
11. Place the jar in a 37 °C incubator. Colonies typically form in 4–7 days.
12. Enumerate the colonies and calculate the CFU/g tissue using the following formula: [(total volume of homogenate) × (# of colonies/volume plated)]/(stomach weight) = CFU/g (see Note 22).
13. Remove colonies of interest from the plates using a separate sterile toothpick for each.
14. Make ¼ in. streaks of the colonies on new blood agar plates, maintaining the additional antibiotics necessary for colonization studies (see Note 23).
15. Place the plates in an Anoxomat jar and evacuate as described above.
16. Place the jar in a 37 °C incubator for 2–4 days allowing the streaks to grow.

17. Remove each streak with a separate sterile cotton swab moistened with Brucella Broth.
18. Take the swab and spread the cells onto ¼ of a new blood agar plate.
19. Place the plates in an Anoxomat jar and evacuate as described above.
20. Place the jar in a 37°C incubator for 1–2 days allowing the lawns to grow.
21. Remove each lawn using a sterile cotton swab.
22. Resuspend the cells in 500 μL of Brucella Broth (see Notes 24 and 25).
23. Take the swab and spread the resuspended cells onto a new blood agar plate.
24. Allow the plate to briefly dry, invert it, and place it in an Anoxomat jar and evacuate as described above.
25. Place the jar in a 37°C incubator for 1–2 days.
26. Place an appropriate amount of freezing media in a 2 mL cryogenic vial (see Notes 26 and 27).
27. Remove lawned *Helicobacter* cells using a sterile cotton swab.
28. Resuspend the cells in the freezing media (see Note 25).
29. Place the cryogenic vial in a –80 °C freezer (see Note 28).

4. Notes

1. All solutions requiring water should be prepared using ultrapure water with a resistance of 18.2 MΩ cm at 25°C.
2. It is helpful to make the agar solution in a 500 mL media bottle, which aids in the mixing of components and the pouring of the solution into petri dishes.
3. The combination of the 1,000× and 200× antibiotic stocks contains a mixture of the common antibiotics found in Skirrow's *Campylobacter* Selective Supplement (Vancomycin, Trimethoprim, and Polymyxin B) (Oxoid, Cambridge, UK, # SR0069) and Dent's *Helicobacter pylori* Selective Supplement (Vancomycin, Trimethoprim, Cefsulodin, and Amphotericin B) (Oxoid, #SR0147). The final concentrations of the antibiotics within the agar are the same as the concentrations when utilizing the supplements with the exception of Amphotericin B, which is at a slightly higher concentration in the preparation presented here.

4. β-Cyclodextrin does not readily go into solution; thus, the DMSO must be added to the β-cyclodextrin and immediately and vigorously vortexed until dissolved.
5. When attempting to isolate colonies from animal colonization studies, it is imperative to incorporate additional antibiotics to isolate *H. pylori* from the normal gut flora found in many rodents.
6. The agar solution must be cooled to prevent the lysis of red blood cells upon addition of the horse blood to the agar.
7. Remove the aliquots of the antibiotic stocks from the −20 °C freezer and allow them to reach room temperature prior to adding them to the agar solution. Cefsulodin often settles out of the solution, so vortex vigorously prior to addition.
8. After the addition of each supplement to the agar solution, swirl the bottle to evenly distribute the supplement and prevent premature solidification of the agar.
9. Final concentrations of the antibiotics are as follows: 5 μg/mL Trimethoprim, 8 μg/mL Amphotericin B, 110 μg/mL Vancomycin Hydrochloride, 5 μg/mL Cefsulodin Sodium Salt, 0.33 μg/mL Polymyxin B Sulfate, 10 μg/mL Nalidixic Acid, and 100 μg/mL Bacitracin.
10. The Bacitracin and Nalidixic Acid solutions do not readily remain in suspension, so vortex vigorously prior to adding them to the bottle of blood agar.
11. β-Cyclodextrin is at a final concentration of 0.2% and the horse blood is 5%.
12. Remove the Defibrinated Horse Blood from the refrigerator and allow it to reach room temperature prior to its addition to the agar solution. This step is important to prevent cool blood from causing the agar to solidify in the bottle.
13. Typically, one 500 mL bottle will provide enough agar for one sleeve containing 20–100 mm petri dishes.
14. To further reduce the number of bubbles in the agar, briefly pass the flame from a Bunsen burner over the surface of the freshly poured agar prior to solidification.
15. To help the plates dry completely after solidification, flip them over so that the lid is down and allow them to sit another 24 h.
16. The plates can typically be used up to 3 weeks after pouring, if they are stored at +4 °C. However, it is important to check for contaminants, since small white colonies tend to form on the agar surface after a month in storage.
17. It is useful to combine equal volumes of 2 sizes of glass beads, 3 and 4 mm (Walter Stern, Inc., Port Washington, NY, #100 C & #100E), in a 250 mL media bottle and autoclave to sterilize.

18. If plating 5 μL or less, it is useful to place 10 μL of Brucella Broth onto the plate immediately prior to adding the bacterial suspension. The additional fluid permits a more thorough distribution of the bacterial cells across the plate.
19. Plating of varying volumes is necessary to obtain colonies separated enough to be counted and removed for expansion. Thus, it may be necessary to further reduce the plating volume to obtain fewer colonies.
20. The beads can be re-used by briefly placing them in a 70% Ethanol solution, washing them thoroughly with de-ionized water, returning them to a 250 mL media bottle, and autoclaving.
21. Alternatively, CampyGen paper sachets (Oxoid, # CN0025) can be used to generate the microaerobic atmosphere in place of the Anoxomat system.
22. To estimate the total homogenate volume, we assume 0.001 g of stomach tissue roughly equals 1 μL. For example, the volume of a homogenate that includes a stomach weighing 0.300 g and 1 mL of Brucella Broth would approximately equal 1,300 μL, since 0.300 g is roughly equivalent to 300 μL. Using the same homogenate, if we obtain a colony count of 50 for a 1 μL plating volume, the CFU/g tissue would equal 2.17×10^5. [(1,300 μL) × (50 colonies/1 μL)]/0.300 g = 2.17×10^5 CFU/g stomach tissue.
23. Take care to keep the streaks separated from each other on the plate to avoid mixing the cells.
24. The amount of Brucella Broth to be used varies from 200 μL to 1 mL based on the density of the growth in the patch.
25. Prior to continuing the culture of *H. pylori* or preparing freezer stocks, it is vital to inspect the morphology of the bacterial cells using a microscope to identify any contaminants. Simply place a 5 μL aliquot on a clean, glass slide, add a coverslip, and view at 400× magnification using phase contrast. Additionally, this step is important to detect the presence of coccoid *Helicobacter* cells, which increase in numbers as cultures age and can affect downstream cultures if a large proportion of the population is in this growth stage.
26. The amount of freezing media to be used varies from 1 to 2 mL based on the density of lawn growth.
27. It is important to use tubes specifically designed for −80°C freezing, such as Nunc cryotube vials (Rochester, NY, #377267).
28. *H. pylori* can be challenging to revive from old freezer stocks, so it is best to maintain freezer stocks for no longer than 2–3 years.

Acknowledgments

We thank Dr. Beth Carpenter for critical reading of the manuscript. Research in the laboratory of D. Scott Merrell is supported by R01AI065529 from NIH, R073PW from USUHS and 300411-7.20-60393 from USMCI. The contents of this work are solely the responsibility of the authors and do not necessarily represent the official views of the NIH or Department of Defense (DoD).

References

1. Kodama M, Murakami K, Nishizono A et al (2004) Animal models for the study of *Helicobacter*-induced gastric carcinoma. J Infect Chemother 10:316–325
2. O'Rourke JL, Lee A (2003) Animal models of *Helicobacter pylori* infection and disease. Microbes Infect 5:741–748

Chapter 6

Successful Culture Techniques for *Helicobacter* Species: Verification of *Helicobacter* Identity Using 16S rRNA Gene Sequence Analysis

Jeannette M. Whitmire and D. Scott Merrell

Abstract

This chapter describes protocols for the verification of putative *Helicobacter* species' identities using 16S rRNA gene sequence analysis.

Key words: *Helicobacter*, *H. pylori*, 16S rRNA gene sequencing

1. Introduction

The advancement of sequencing techniques has led to the publication of the complete genome sequences for several *Helicobacter* isolates (1–9). The availability of these complete genomic sequences enhances researchers' abilities to positively identify potential *Helicobacter* isolates. In particular, sequencing of the 16S rRNA gene from strains of interest allows comparison to the known sequences of various *Helicobacter* species for verification (1–25). Protocols are presented to describe methods for the verification of putative *Helicobacter* species' identities using 16S rRNA gene sequencing.

JeanMarie Houghton (ed.), *Helicobacter Species: Methods and Protocols*, Methods in Molecular Biology, vol. 921,
DOI 10.1007/978-1-62703-005-2_6, © Springer Science+Business Media, LLC 2012

2. Materials

1. Phusion High Fidelity Enzyme Kit (New England BioLabs, Ipswich, MA, #F-540L).
2. DNase/RNase-Free Distilled Water.
3. MinElute PCR Purification Kit (Qiagen, Valencia, CA, #28006).
4. BigDye Terminator v3.1 (Applied Biosystems, Carlsbad, CA, #4337458).
5. Performa DTR Gel Filtration Cartridges (Edge Biosystems, Gaithersburg, MD, #42453).

3. Methods

1. Isolate DNA from the strain of interest using a commercially available kit, such as the QIAamp DNA Mini Kit (Qiagen, #51306) or NucleoSpin Tissue Kit (Marcherey-Nagel, Inc., Bethlehem, PA, #740952.250).
2. Set up a PCR reaction with 50 ng of genomic DNA from the strain using 1 μM each of both 8F and 1492R primers, 1 μM of dNTPs, 1–2 Units of Phusion enzyme, and balance with the appropriate amount of buffer and nuclease-free water (see Note 1).
3. Amplify the 16S rRNA gene using the following thermal cycling conditions: Initial Denaturation at 98°C for 2 min; 30 cycles with a Denaturation at 98°C for 5 s, Annealing at 52°C for 20 s, and Extension at 72°C for 20 s; and Final Extension at 72°C for 5 min.
4. Clean up the PCR reaction using a MinElute PCR Purification Kit according to the directions provided by the manufacturer.
5. Determine the concentration of DNA from the cleaned PCR reaction.
6. Set up separate sequencing reactions: one using primer 515F and the other with primer 907R (see Note 1). Use 1 μL BigDye, 100–150 ng PCR product, 2 μL 5× Buffer, 1 μL primer, and balance to 10 μL with nuclease-free water.
7. Perform the sequencing reactions with the following cycling conditions: Initial Denaturation at 96°C for 1 min; 15 cycles with a Denaturation at 96°C for 10 s, Annealing at 55°C for 5 s, and Extension at 60°C for 1.5 min; 5 cycles with a Denaturation at 96°C for 10 s, Annealing at 55°C for 5 s, and Extension at 60°C for 1.75 min; and 5 cycles with a

Denaturation at 96°C for 30 s, Annealing at 55°C for 5 s, and Extension at 60°C for 2 min.

8. Clean up the sequencing reaction with a Performa DTR cartridge according to the manufacturer's directions.
9. Run the reaction on a 3130XL Genetic Analyzer (Applied Biosystems) to obtain the sequence.
10. Compare the sequence to published *Helicobacter* 16S rRNA sequences to verify the species (1–19, 23, 24).

4. Note

1. Primer Sequences (5′-3′): 8F-AGAGTTTGATCCTGGCTCAG, 1492R-GGTTACCTTGTTACGACTT (26), 515F-GTGCAAGCM GCCGCGGTA (27), 907R-CCGTCAATTCCTTTRAGTTT (28).

References

1. Alm RA, Ling LS, Moir DT et al (1999) Genomic-sequence comparison of two unrelated isolates of the human gastric pathogen *Helicobacter pylori*. Nature 397:176–180
2. Baltrus DA, Amieva MR, Covacci A et al (2009) The complete genome sequence of *Helicobacter pylori* strain G27. J Bacteriol 191:447–448
3. Farnbacher M, Jahns T, Willrodt D et al (2010) Sequencing, annotation, and comparative genome analysis of the gerbil-adapted *Helicobacter pylori* strain B8. BMC Genomics 11:335
4. Fischer W, Windhager L, Rohrer S et al (2010) Strain-specific genes of *Helicobacter pylori*: genome evolution driven by a novel type IV secretion system and genomic island transfer. Nucleic Acids Res 38:6089–6101
5. Giannakis M, Chen SL, Karam SM et al (2008) *Helicobacter pylori* evolution during progression from chronic atrophic gastritis to gastric cancer and its impact on gastric stem cells. Proc Natl Acad Sci U S A 105:4358–4363
6. Oh JD, Kling-Backhed H, Giannakis M et al (2006) The complete genome sequence of a chronic atrophic gastritis *Helicobacter pylori* strain: evolution during disease progression. Proc Natl Acad Sci U S A 103:9999–10004
7. Suerbaum S, Josenhans C, Sterzenbach T et al (2003) The complete genome sequence of the carcinogenic bacterium *Helicobacter hepaticus*. Proc Natl Acad Sci U S A 100:7901–7906
8. Thiberge JM, Boursaux-Eude C, Lehours P et al (2010) From array-based hybridization of *Helicobacter pylori* isolates to the complete genome sequence of an isolate associated with MALT lymphoma. BMC Genomics 11: 368–379
9. Tomb JF, White O, Kerlavage AR et al (1997) The complete genome sequence of the gastric pathogen *Helicobacter pylori*. Nature 388: 539–547
10. Fox JG, Yan LL, Dewhirst FE et al (1995) *Helicobacter bilis* sp. nov., a novel *Helicobacter* species isolated from bile, livers, and intestines of aged, inbred mice. J Clin Microbiol 33: 445–454
11. Fox JG, Dewhirst FE, Tully JG et al (1994) *Helicobacter hepaticus* sp. nov., a microaerophilic bacterium isolated from livers and intestinal mucosal scrapings from mice. J Clin Microbiol 32:1238–1245
12. Battles JK, Williamson JC, Pike KM et al (1995) Diagnostic assay for *Helicobacter hepaticus* based on nucleotide sequence of its 16S rRNA gene. J Clin Microbiol 33: 1344–1347
13. Cattoli G, van Vugt R, Zanoni RG et al (1999) Occurrence and characterization of gastric *Helicobacter* spp. in naturally infected dogs. Vet Microbiol 70:239–250
14. Dewhirst FE, Shen Z, Scimeca MS et al (2005) Discordant 16S and 23S rRNA gene phylogenies for the genus *Helicobacter*: implications

for phylogenetic inference and systematics. J Bacteriol 187:6106–6118
15. Eaton KA, Dewhirst FE, Paster BJ et al (1996) Prevalence and varieties of *Helicobacter* species in dogs from random sources and pet dogs: animal and public health implications. J Clin Microbiol 34:3165–3170
16. Eaton KA, Dewhirst FE, Radin MJ et al (1993) *Helicobacter acinonyx* sp. nov., isolated from cheetahs with gastritis. Int J Syst Bacteriol 43:99–106
17. Eckloff BW, Podzorski RP, Kline BC et al (1994) A comparison of 16S ribosomal DNA sequences from five isolates of *Helicobacter pylori*. Int J Syst Bacteriol 44:320–323
18. Fox JG, Dewhirst FE, Shen Z et al (1998) Hepatic *Helicobacter* species identified in bile and gallbladder tissue from Chileans with chronic cholecystitis. Gastroenterology 114:755–763
19. Hanninen ML, Karenlampi RI, Koort JM et al (2005) Extension of the species *Helicobacter bilis* to include the reference strains of *Helicobacter* sp. flexispira taxa 2, 3 and 8 and Finnish canine and feline flexispira strains. Int J Syst Evol Microbiol 55:891–898
20. Kolbert CP, Persing DH (1999) Ribosomal DNA sequencing as a tool for identification of bacterial pathogens. Curr Opin Microbiol 2:299–305
21. Liu H, Rahman A, Semino-Mora C et al (2008) Specific and sensitive detection of *H. pylori* in biological specimens by real-time RT-PCR and in situ hybridization. PLoS One 3:e2689
22. Mahler M, Bedigian HG, Burgett BL et al (1998) Comparison of four diagnostic methods for detection of *Helicobacter* species in laboratory mice. Lab Anim Sci 48:85–91
23. O'Rourke JL, Solnick JV, Neilan BA et al (2004) Description of "*Candidatus Helicobacter heilmannii*" based on DNA sequence analysis of 16S rRNA and urease genes. Int J Syst Evol Microbiol 54:2203–2211
24. Paster BJ, Lee A, Fox JG et al (1991) Phylogeny of *Helicobacter felis* sp. nov., *Helicobacter mustelae*, and related bacteria. Int J Syst Bacteriol 41:31–38
25. Recordati C, Gualdi V, Craven M et al (2009) Spatial distribution of *Helicobacter* spp. in the gastrointestinal tract of dogs. Helicobacter 14:180–191
26. Reed DW, Fujita Y, Delwiche ME et al (2002) Microbial communities from methane hydrate-bearing deep marine sediments in a forearc basin. Appl Environ Microbiol 68(8):3759–3770
27. Miteva VI, Sheridan PP, Brenchley JE (2004) Phylogenetic and physiological diversity of microorganisms isolated from a deep greenland glacier ice core. Appl Environ Microbiol 70(1):202–213
28. Reysenbach AL, Longnecker K, Kirshtein J (2000) Novel bacterial and archaeal lineages from an in situ growth chamber deployed at a Mid-Atlantic Ridge hydrothermal vent. Appl Environ Microbiol 66(9):3798–3806

Chapter 7

The *Helicobacter pylori cag* Pathogenicity Island

Jennifer M. Noto and Richard M. Peek Jr.

Abstract

The *cag* pathogenicity island is a well-characterized virulence determinant. It is composed of 32 genes that encode a type IV bacterial secretion system and is linked with a more severe clinical outcome. The following chapters will explore the manipulation of bacterial factors in order to understand their role in gastric mucosal disease.

Key words: *Helicobacter pylori*, *cag* pathogenicity island

1. The *Helicobacter pylori cag* Pathogenicity Island

H. pylori strains exhibit a high degree of genetic heterogeneity due to genomic rearrangements, point mutations, gene insertions, and/or deletions (1–4). Genetically unique variants of a single strain are present within the stomachs of each human host, and the genetic composition of these populations can evolve over time (5). The identification of bacterial factors clearly associated with disease outcomes has been hindered because of this high level of genetic diversity; however, specific loci have been identified that augment the risk for carcinogenesis. The *cag* pathogenicity island (*cag* PAI) is a 40 kb DNA insertion element, containing approximately 32 genes that encode a bacterial type IV secretion system (4, 6–8). The *cag* PAI is a well-characterized *H. pylori* virulence determinant that is present in approximately 60–70% of Western *H. pylori* strains and virtually 100% of East-Asian *H. pylori* strains (4, 6–8). Although all *H. pylori* strains induce gastritis, strains that harbor the *cag* PAI (cag^{+}) augment the risk for severe gastritis, atrophy, dysplasia, and gastric adenocarcinoma compared to strains that lack the *cag* island

JeanMarie Houghton (ed.), *Helicobacter Species: Methods and Protocols*, Methods in Molecular Biology, vol. 921,
DOI 10.1007/978-1-62703-005-2_7,

(*cag⁻*) (9–20). *H. pylori cag⁻* strains are found predominantly in the mucus gel layer while *cag⁺* strains are found adjacent to or adherent to gastric epithelial cells, indicating that *cag* genotype influences the topography of colonization within the stomach (21).

The *cag* type IV secretion system allows for the delivery of bacterial effector molecules into host gastric epithelial cells. One of these proteins, CagE, is a structural component of the functional type IV secretion system and inactivation of this gene product abrogates delivery of *H. pylori* proteins into host cells. Another component of the secretion system, CagL, functions as a specialized bacterial adhesin that binds to and activates $\alpha_5\beta_1$ integrin receptors, triggering the delivery of bacterial effector molecules into the cytoplasm of host cells (22). CagL bridges the type IV secretion system to $\alpha_5\beta_1$ integrins on target cells and activates host focal adhesion kinase (FAK) and Src (Fig. 1). In addition to $\alpha_5\beta_1$ integrin, CagL can also bind integrin and fibronectin, although the downstream consequences of binding to these receptors remains undefined. Recently, additional Cag proteins (CagA, CagI, CagY)

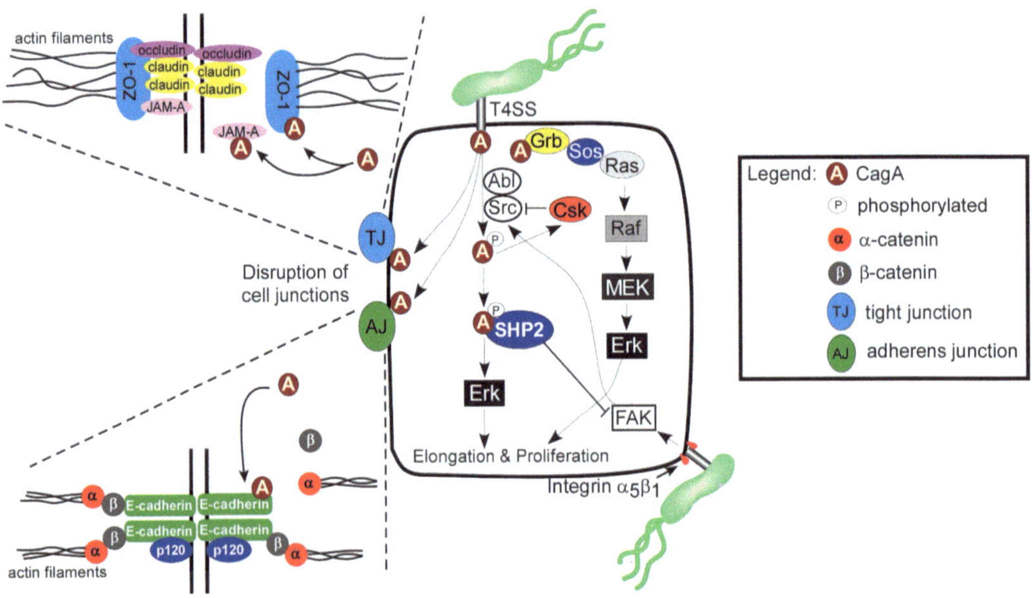

Fig. 1. Molecular signaling alterations induced by intracellular delivery of CagA. Translocation of CagA by the *H. pylori cag* type IV secretion system leads to activation of host signaling pathways that promote epithelial responses with carcinogenic potential. CagA is phosphorylated by Src and Abl kinases, which is followed by a decrease in levels of phosphorylated-CagA via the inhibitory kinase c-src kinase (Csk). Phosphorylated CagA activates SHP2 and Erk leading to morphological changes, such as cellular elongation. Additionally, the interaction between phosphorylated-CagA and SHP2 results in inactivation of focal adhesion kinase (FAK), which can activate Src. Unphosphorylated CagA also leads to changes in epithelial cell motility and proliferation through binding Grb/Sos/Ras and activation of the Raf/MEK/Erk pathway. Unmodified CagA can also associate with the tight junction proteins ZO-1 and JAM-A as well as the adherens junction protein E-cadherin, leading to dysregulated junctional complexes.

have been shown to bind β_1 integrin and induce conformational changes of integrin heterodimers, permitting translocation of bacterial effectors (23).

2. CagA

The terminal gene product of the *cag* island, CagA, is translocated into host cells by the *cag* type IV secretion system following bacterial attachment (24). Transgenic mice that overexpress CagA develop gastric epithelial cell hyperproliferation and gastric adenocarcinoma (25), implicating this molecule as a bacterial oncoprotein. CagA is a 120–140 kD protein that contains tyrosine phosphorylation motifs (glutamate-proline-isoleucine-tyrosine-alanine, EPIYA) within the carboxyl-terminal variable region of the protein (26). To date, four distinct EPIYA motifs have been identified within the carboxyl-terminal polymorphic region of CagA. These motifs are designated EPIYA -A, -B, -C, or -D and are distinguished by the amino acid sequences flanking the EPIYA motif (27–29). Most variants of CagA throughout the world contain EPIYA-A and -B motifs, while EPIYA-C motifs are predominantly found in strains from Western countries (Europe, North America, and Australia). The number of EPIYA-C sites can vary; however, most CagA proteins contain a single EPIYA-C site (A-B-C-type). The EPIYA-A and -B motifs are phosphorylated to a lesser extent than EPIYA-C motifs and in Western strains an increased number of CagA EPIYA-C motifs correlates with increased gastric cancer risk (30, 31). EPIYA-D motifs are found almost exclusively in East-Asian *H. pylori* strains (Japan, Korea, and China) and are phosphorylated to a greater extent than all other EPIYA motifs (27). *H. pylori* strains containing EPIYA-D motifs induce significantly higher levels of IL-8 release from gastric epithelial cells compared to strains harboring Western A-B-C-type CagA (27, 32). Thus, the majority of *cag*$^+$ Western strains express A-B-C-type CagA, and the number of EPIYA-C regions may vary between 1 and 3 repeated copies among different strains, while East-Asian strains typically express A-B-D-type CagA (27).

3. CagA Phosphorylation-Dependent Perturbation of Host Cell Signaling

Following its injection into epithelial cells, CagA undergoes tyrosine phosphorylation at EPIYA motifs by members of the Abl and Src family of kinases (24, 33–37) (Fig. 1). Intracellular, phosphorylated-CagA, in turn, activates a eukaryotic phosphatase (SHP-2) and extracellular signal-regulated kinase 1 and 2 (Erk1/2), leading to cell scattering, robust actin reorganization, and other morphologic changes reminiscent of unrestrained stimulation by growth factors (24, 26, 33–41). Specifically, CagA transfection studies have demonstrated that phosphorylated-CagA-SHP-2 interactions contribute to cytoskeletal rearrangements and cell elongation by activation of the Erk-signaling pathway (38). East-Asian A-B-D-type CagA exhibits a higher binding affinity for SHP-2 than Western A-B-C-type CagA and, therefore, induces a more robust morphologic response by gastric epithelial cells (39).

H. pylori CagA proteins tightly and specifically regulate the activity of Src and Abl family kinases in a time-dependent manner. Src is activated during the initial stages of infection and is then rapidly inactivated, while Abl is continuously activated by *H. pylori* with enhanced activities at later time points, supporting a model of successive phosphorylation of CagA by Src and Abl family kinases (42). Phosphorylated-CagA can also inhibit Src via recruitment of C-terminal Src kinase (Csk), a negative-regulator of Src that acts rapidly to initiate a negative feedback loop to downregulate Src signaling (41, 43) (Fig. 1). As Src is the primary kinase activated by CagA, inhibition of Src by phosphorylated-CagA generates a negative feedback loop that carefully regulates the amount of intracellular, phosphorylated-CagA. The catalytic activity of Src is inhibited by phosphorylated-CagA, leading to tyrosine dephosphorylation of the actin-binding proteins cortactin, ezrin, and vinculin, which ultimately results in cellular rearrangements and elongation (43–45).

In AGS human gastric epithelial cells, translocation and subsequent phosphorylation of CagA results in cell elongation and scattering, known as the "hummingbird" phenotype (35, 46). In this cell line, interactions between phosphorylated-CagA and SHP-2 increase the duration of Erk activation in a Ras- and PI3K-independent manner, resulting in cell elongation (38). The interaction between phosphorylated-CagA and SHP-2 also results in dephosphorylation and inactivation of FAK, which again leads to morphologic aberrations (47) (Fig. 1).

4. CagA Phosphorylation-Independent Perturbation of Host Cell Signaling

Non-phosphorylated CagA also exerts numerous effects within gastric epithelial cells that contribute to pathogenesis. CagA translocation, but not phosphorylation, leads to disruption of apical–junctional complexes (Fig. 1). Non-phosphorylated CagA associates with the epithelial tight-junction scaffolding protein zona occludens 1 (ZO-1) and the transmembrane protein junctional adhesion molecule A (JAM-A), leading to nascent but incomplete assembly of tight junctions (TJ) at ectopic sites of bacterial attachment (48).

In addition, non-phosphorylated CagA disrupts adherens junctions (AJ) leading to aberrant activation of β-catenin and an overall loss of barrier function and cellular polarity (48–53), alterations that play an important role in carcinogenesis. Non-phosphorylated CagA interacts with the cell adhesion protein E-cadherin, the hepatocyte growth factor receptor c-Met, the phospholipase PLC-γ, and the adaptor protein Grb2 (51, 52, 54, 55), which leads to pro-inflammatory and mitogenic responses, disruption of cellular junctions, and loss of cellular polarity. Recently, non-phosphorylated CagA was shown to directly bind PAR1b/MARK2, a central regulator of cell polarity, and inhibit its kinase activity. This interaction induced the dysregulation of mitotic spindle formation, promoting a loss of cellular polarity (52, 56, 57). These events were dependent on conserved 16 amino acid repeat motifs embedded within the 3 terminus of CagA and which are known as CagA multimerization (CM) sites (58), the conserved repeat responsible for phosphorylation-independent activity (CRPIA) (59), or the PAR1b/MARK2 kinase inhibitor (MKI) (60) motifs. These motifs, which vary in number among *H. pylori* strains, bind PAR1b/MARK2 and mediate homodimerization of CagA, conferring heightened SHP-2 binding and activation. A recent co-crystallography analysis of the CagA-PAR1b/MARK2 complex demonstrated that the PAR1b/MARK2-binding site resides in the initial 14 amino acids of the CagA CM motif, and binding leads to inhibition of PAR1b/MARK2 kinase activity (60).

5. Peptidoglycan

Another consequence of *H. pylori cag* pathogenicity island-mediated host cell contact is the production of pro-inflammatory cytokines. In certain *H. pylori* strains, CagA can induce IL-8 expression via NF-κB activation (61–63); however, the ability of CagA to induce IL-8 expression is not universal across all *cag* PAI-bearing strains (64–67). In addition to CagA, *H. pylori* peptidoglycan (PGN) is delivered into host cells via the *cag* type IV secretion system and outer membrane vesicles (OMV) (68), where they are sensed by the nucleotide-binding oligomerization domain 1 (NOD1), an intracytoplasmic pathogen pattern-recognition molecule (69, 70) (Fig. 2). NOD1 activation by *H. pylori* peptidoglycan stimulates the production of pro-inflammatory cytokines MIP-2, β-defensin, and IL-8 through induction of host cells signaling molecules, nuclear factor κB (NF-κB), p38, and Erk (70, 71). Furthermore, NOD1 activation by *H. pylori* peptidoglycan also regulates the production of type I interferon (IFN), which likely affects Th1 cell differentiation (72). NOD1-deficient mice develop an attenuated mucosal cytokine response following infection with

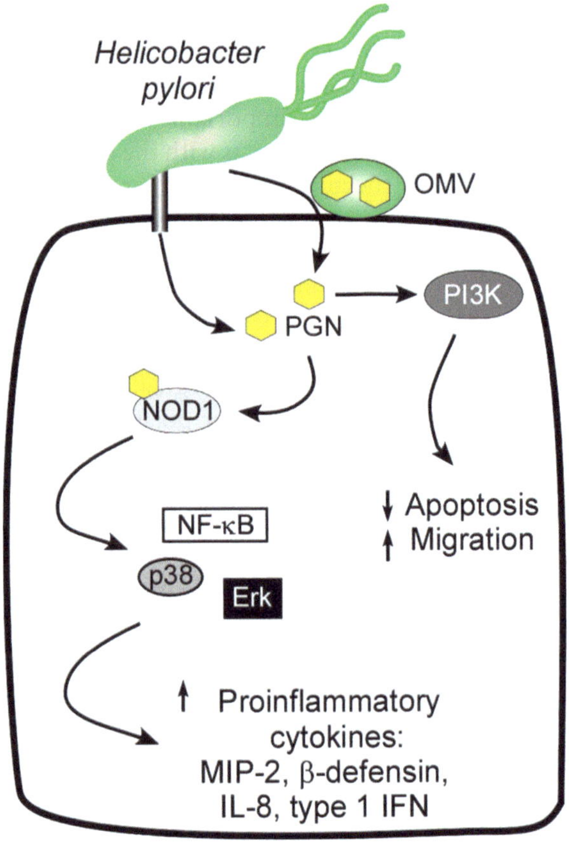

Fig. 2. Molecular signaling alterations induced by intracellular delivery of peptidoglycan. In addition to CagA, the *H. pylori cag* type IV secretion system can deliver peptidoglycan (PGN) into host cells. Another mechanism of PGN delivery is via outer membrane vesicles (OMV). Delivery of PGN results in activation of the intracellular receptor nucleotide oligomerization domain 1 (NOD1) and triggers multiple signaling pathways that culminate in NF-κB activation and subsequent production of inflammatory and immune effectors, such as IL-8 and Type 1 IFN. Further, PGN can also activate PI3K, leading to decreased levels of apoptosis and increased cell migration.

H. pylori cag$^+$ strains (73), implicating peptidoglycan-NOD1 signaling as an important mediator of *H. pylori* pathogenesis.

Delivery of peptidoglycan components into host cells induces additional epithelial responses with carcinogenic potential, such as activation of PI3K and cell migration (Fig. 2). The *H. pylori* gene *slt* encodes a soluble lytic transglycosylase that is required for peptidoglycan turnover and release (70), thereby regulating the amount of peptidoglycan translocated into host cells. Inactivation of *slt* has now been shown to inhibit *H. pylori*-induced PI3K signaling and cell migration (74). The protein encoded by the *H. pylori* gene *HP0310* deacetylates *N*-acetylglucosamine peptidoglycan residues and is required for peptidoglycan synthesis (75). Loss of *HP0310*, leading to decreased peptidoglycan production, reciprocally augments delivery of the other major *cag* secretion system substrate, CagA, into host cells, suggesting that functional interactions occur between *H. pylori* translocated effector molecules (76). Further, *H. pylori* peptidoglycan deacetylation by HP0310 is an important mechanism for mitigating host immune detection, which facilitates bacterial persistence and colonization (77). In total, these findings indicate that contact between *cag*[+] strains and host cells activates multiple signaling pathways that regulate oncogenic cellular responses, which may heighten the risk for transformation, particularly over prolonged periods of *H. pylori* infection.

References

1. Alm RA, Bina J, Andrews BM, Doig P, Hancock RE, Trust TJ (2000) Comparative genomics of *Helicobacter pylori*: analysis of the outer membrane protein families. Infect Immun 68:4155–4168
2. Go MF, Kapur V, Graham DY, Musser JM (1996) Population genetic analysis of *Helicobacter pylori* by multilocus enzyme electrophoresis: extensive allelic diversity and recombinational population structure. J Bacteriol 178:3934–3938
3. Salama N, Guillemin K, McDaniel TK, Sherlock G, Tompkins L, Falkow S (2000) A whole-genome microarray reveals genetic diversity among *Helicobacter pylori* strains. Proc Natl Acad Sci USA 97:14668–14673
4. Tomb JF, White O, Kerlavage AR, Clayton RA, Sutton GG, Fleischmann RD et al (1997) The complete genome sequence of the gastric pathogen *Helicobacter pylori*. Nature 388:539–547
5. Israel DA, Salama N, Krishna U, Rieger UM, Atherton JC, Falkow S et al (2001) *Helicobacter pylori* genetic diversity within the gastric niche of a single human host. Proc Natl Acad Sci USA 98:14625–14630
6. Akopyants NS, Clifton SW, Kersulyte D, Crabtree JE, Youree BE, Reece CA et al (1998) Analyses of the *cag* pathogenicity island of *Helicobacter pylori*. Mol Microbiol 28:37–53
7. Alm RA, Ling LS, Moir DT, King BL, Brown ED, Doig PC et al (1999) Genomic-sequence comparison of two unrelated isolates of the human gastric pathogen *Helicobacter pylori*. Nature 397:176–180
8. Censini S, Lange C, Xiang Z, Crabtree JE, Ghiara P, Borodovsky M et al (1996) *Cag*, a pathogenicity island of *Helicobacter pylori*, encodes type I specific and disease-associated virulence factors. Proc Natl Acad Sci USA 93:14648–14653
9. Blaser MJ, Perez-Perez GI, Kleanthous H, Cover TL, Peek RM, Chyou PH et al (1995) Infection with *Helicobacter pylori* strains possessing *cagA* is associated with an increased risk of developing adenocarcinoma of the stomach. Cancer Res 55:2111–2115
10. Cover TL, Dooley CP, Blaser MJ (1990) Characterization of and human serologic response to proteins in *Helicobacter pylori* broth culture supernatants with vacuolizing cytotoxin activity. Infect Immun 58:603–610
11. Crabtree JE, Taylor JD, Wyatt JI, Heatley RV, Shallcross TM, Tompkins DS et al (1991) Mucosal IgA recognition of *Helicobacter pylori* 120 kDa protein, peptic ulceration, and gastric pathology. Lancet 338:332–335
12. Crabtree JE, Wyatt JI, Sobala GM, Miller G, Tompkins DS, Primrose JN et al (1993) Systemic and mucosal humoral responses to *Helicobacter pylori* in gastric cancer. Gut 34:1339–1343
13. Kuipers EJ, Perez-Perez GI, Meuwissen SG, Blaser MJ (1995) *Helicobacter pylori* and atrophic gastritis: importance of the *cagA* status. J Natl Cancer Inst 87:1777–1780
14. Parsonnet J, Replogle M, Yang S, Hiatt R (1997) Seroprevalence of CagA-positive strains among *Helicobacter pylori*-infected, healthy young adults. J Infect Dis 175:1240–1242
15. Peek RM Jr, Miller GG, Tham KT, Perez-Perez GI, Zhao X, Atherton JC et al (1995) Heightened inflammatory response and cytokine expression in vivo to *cagA*[+] *Helicobacter pylori* strains. Lab Invest 73:760–770
16. Queiroz DM, Mendes EN, Rocha GA, Oliveira AM, Oliveira CA, Magalhaes PP et al (1998) *CagA*-positive *Helicobacter pylori* and risk for developing gastric carcinoma in Brazil. Int J Cancer 78:135–139

17. Rudi J, Kolb C, Maiwald M, Zuna I, von Herbay A, Galle PR et al (1997) Serum antibodies against *Helicobacter pylori* proteins VacA and CagA are associated with increased risk for gastric adenocarcinoma. Dig Dis Sci 42:1652–1659
18. Shimoyama T, Fukuda S, Tanaka M, Mikami T, Munakata A, Crabtree JE (1998) CagA seropositivity associated with development of gastric cancer in a Japanese population. J Clin Pathol 51:225–228
19. Torres J, Perez-Perez GI, Leal-Herrera Y, Munoz O (1998) Infection with CagA+ *Helicobacter pylori* strains as a possible predictor of risk in the development of gastric adenocarcinoma in Mexico. Int J Cancer 78:298–300
20. Vorobjova T, Nilsson I, Kull K, Maaroos HI, Covacci A, Wadstrom T et al (1998) CagA protein seropositivity in a random sample of adult population and gastric cancer patients in Estonia. Eur J Gastroenterol Hepatol 10:41–46
21. Camorlinga-Ponce M, Romo C, Gonzalez-Valencia G, Munoz O, Torres J (2004) Topographical localisation of *cagA* positive and *cagA* negative *Helicobacter pylori* strains in the gastric mucosa; an in situ hybridisation study. J Clin Pathol 57:822–828
22. Kwok T, Zabler D, Urman S, Rohde M, Hartig R, Wessler S et al (2007) *Helicobacter* exploits integrin for type IV secretion and kinase activation. Nature 449:862–866
23. Jimenez-Soto LF, Kutter S, Sewald X, Ertl C, Weiss E, Kapp U et al (2009) *Helicobacter pylori* type IV secretion apparatus exploits beta1 integrin in a novel RGD-independent manner. PLoS Pathogens 5:e1000684
24. Odenbreit S, Puls J, Sedlmaier B, Gerland E, Fischer W, Haas R (2000) Translocation of *Helicobacter pylori* CagA into gastric epithelial cells by type IV secretion. Science 287:1497–1500
25. Ohnishi N, Yuasa H, Tanaka S, Sawa H, Miura M, Matsui A et al (2008) Transgenic expression of *Helicobacter pylori* CagA induces gastrointestinal and hematopoietic neoplasms in mouse. Proc Natl Acad Sci USA 105:1003–1008
26. Stein M, Bagnoli F, Halenbeck R, Rappuoli R, Fantl WJ, Covacci A (2002) c-Src/Lyn kinases activate *Helicobacter pylori* CagA through tyrosine phosphorylation of the EPIYA motifs. Mol Microbiol 43:971–980
27. Hatakeyama M (2004) Oncogenic mechanisms of the *Helicobacter pylori* CagA protein. Nat Rev Cancer 4:688–694
28. Higashi H, Yokoyama K, Fujii Y, Ren S, Yuasa H, Saadat I et al (2005) EPIYA motif is a membrane-targeting signal of *Helicobacter pylori* virulence factor CagA in mammalian cells. J Biol Chem 280:23130–23137
29. Naito M, Yamazaki T, Tsutsumi R, Higashi H, Onoe K, Yamazaki S et al (2006) Influence of EPIYA-repeat polymorphism on the phosphorylation-dependent biological activity of *Helicobacter pylori* CagA. Gastroenterology 130:1181–1190
30. Basso D, Zambon CF, Letley DP, Stranges A, Marchet A, Rhead JL et al (2008) Clinical relevance of *Helicobacter pylori cagA* and *vacA* gene polymorphisms. Gastroenterology 135:91–99
31. Schneider N, Krishna U, Romero-Gallo J, Israel DA, Piazuelo MB, Camargo MC et al (2009) Role of *Helicobacter pylori* CagA molecular variations in induction of host phenotypes with carcinogenic potential. J Infect Dis 199:1218–1221
32. Argent RH, Hale JL, El-Omar EM, Atherton JC (2008) Differences in *Helicobacter pylori* CagA tyrosine phosphorylation motif patterns between western and East Asian strains, and influences on interleukin-8 secretion. J Med Microbiol 57:1062–1067
33. Asahi M, Azuma T, Ito S, Ito Y, Suto H, Nagai Y et al (2000) *Helicobacter pylori* CagA protein can be tyrosine phosphorylated in gastric epithelial cells. J Exp Med 191:593–602
34. Backert S, Ziska E, Brinkmann V, Zimny-Arndt U, Fauconnier A, Jungblut PR et al (2000) Translocation of the *Helicobacter pylori* CagA protein in gastric epithelial cells by a type IV secretion apparatus. Cell Microbiol 2:155–164
35. Segal ED, Cha J, Lo J, Falkow S, Tompkins LS (1999) Altered states: involvement of phosphorylated CagA in the induction of host cellular growth changes by *Helicobacter pylori*. Proc Natl Acad Sci USA 96:14559–14564
36. Selbach M, Moese S, Hauck CR, Meyer TF, Backert S (2002) Src is the kinase of the *Helicobacter pylori* CagA protein in vitro and in vivo. J Biol Chem 277:6775–6778
37. Stein M, Rappuoli R, Covacci A (2000) Tyrosine phosphorylation of the *Helicobacter pylori* CagA antigen after cag-driven host cell translocation. Proc Natl Acad Sci USA 97:1263–1268
38. Higashi H, Nakaya A, Tsutsumi R, Yokoyama K, Fujii Y, Ishikawa S et al (2004) *Helicobacter pylori* CagA induces Ras-independent morphogenetic response through SHP-2

recruitment and activation. J Biol Chem 279:17205–17216

39. Higashi H, Tsutsumi R, Fujita A, Yamazaki S, Asaka M, Azuma T et al (2002) Biological activity of the *Helicobacter pylori* virulence factor CagA is determined by variation in the tyrosine phosphorylation sites. Proc Natl Acad Sci USA 99:14428–14433
40. Higashi H, Tsutsumi R, Muto S, Sugiyama T, Azuma T, Asaka M et al (2002) SHP-2 tyrosine phosphatase as an intracellular target of *Helicobacter pylori* CagA protein. Science 295:683–686
41. Tsutsumi R, Higashi H, Higuchi M, Okada M, Hatakeyama M (2002) Attenuation of *Helicobacter pylori* CagA-SHP-2 signaling by interaction between CagA and C-terminal Src kinase. J Biol Chem 21:21
42. Tammer I, Brandt S, Hartig R, Konig W, Backert S (2007) Activation of Abl by *Helicobacter pylori*: a novel kinase for CagA and crucial mediator of host cell scattering. Gastroenterology 132:1309–1319
43. Selbach M, Moese S, Hurwitz R, Hauck CR, Meyer TF, Backert S (2003) The *Helicobacter pylori* CagA protein induces cortactin dephosphorylation and actin rearrangement by c-Src inactivation. EMBO J 22:515–528
44. Moese S, Selbach M, Brinkmann V, Karlas A, Haimovich B, Backert S et al (2007) The *Helicobacter pylori* CagA protein disrupts matrix adhesion of gastric epithelial cells by dephosphorylation of vinculin. Cell Microbiol 9:1148–1161
45. Selbach M, Moese S, Backert S, Jungblut PR, Meyer TF (2004) The *Helicobacter pylori* CagA protein induces tyrosine dephosphorylation of ezrin. Proteomics 4:2961–2968
46. Moese S, Selbach M, Kwok T, Brinkmann V, Konig W, Meyer TF et al (2004) *Helicobacter pylori* induces AGS cell motility and elongation via independent signaling pathways. Infect Immun 72:3646–3649
47. Tsutsumi R, Takahashi A, Azuma T, Higashi H, Hatakeyama M (2006) Focal adhesion kinase is a substrate and downstream effector of SHP-2 complexed with *Helicobacter pylori* CagA. Mol Cell Biol 26:261–276
48. Amieva MR, Vogelmann R, Covacci A, Tompkins LS, Nelson WJ, Falkow S (2003) Disruption of the epithelial apical-junctional complex by *Helicobacter pylori* CagA. Science 300:1430–1434
49. Bagnoli F, Buti L, Tompkins L, Covacci A, Amieva MR (2005) *Helicobacter pylori* CagA induces a transition from polarized to invasive phenotypes in MDCK cells. Proc Natl Acad Sci USA 102:16339–16344
50. Franco AT, Israel DA, Washington MK, Krishna U, Fox JG, Rogers AB et al (2005) Activation of beta-catenin by carcinogenic *Helicobacter pylori*. Proc Natl Acad Sci USA 102:10646–10651
51. Murata-Kamiya N, Kurashima Y, Teishikata Y, Yamahashi Y, Saito Y, Higashi H et al (2007) *Helicobacter pylori* CagA interacts with E-cadherin and deregulates the beta-catenin signal that promotes intestinal transdifferentiation in gastric epithelial cells. Oncogene 26:4617–4626
52. Saadat I, Higashi H, Obuse C, Umeda M, Murata-Kamiya N, Saito Y et al (2007) *Helicobacter pylori* CagA targets PAR1/MARK kinase to disrupt epithelial cell polarity. Nature 447:330–333
53. Suzuki M, Mimuro H, Suzuki T, Park M, Yamamoto T, Sasakawa C (2005) Interaction of CagA with Crk plays an important role in *Helicobacter pylori*-induced loss of gastric epithelial cell adhesion. J Exp Med 202:1235–1247
54. Churin Y, Al-Ghoul L, Kepp O, Meyer TF, Birchmeier W, Naumann M (2003) *Helicobacter pylori* CagA protein targets the c-Met receptor and enhances the mitogenic response. J Cell Biol 161:249–255
55. Mimuro H, Suzuki T, Tanaka J, Asahi M, Haas R, Sasakawa C (2002) Grb2 is a key mediator of *Helicobacter pylori* CagA protein activities. Mol Cell 10:745–755
56. Lu H, Murata-Kamiya N, Saito Y, Hatakeyama M (2009) Role of partitioning-defective 1/microtubule affinity-regulating kinases in the morphogenetic activity of *Helicobacter pylori* CagA. J Biol Chem 284:23024–23036
57. Umeda M, Murata-Kamiya N, Saito Y, Ohba Y, Takahashi M, Hatakeyama M (2009) *Helicobacter pylori* CagA causes mitotic impairment and induces chromosomal instability. J Biol Chem 284:22166–22172
58. Kurashima Y, Murata-Kamiya N, Kikuchi K, Higashi H, Azuma T, Kondo S et al (2008) Deregulation of beta-catenin signal by *Helicobacter pylori* CagA requires the CagA-multimerization sequence. Int J Cancer 122:823–831
59. Suzuki M, Mimuro H, Kiga K, Fukumatsu M, Ishijima N, Morikawa H et al (2009) *Helicobacter pylori* CagA phosphorylation-independent function in epithelial proliferation and inflammation. Cell Host Microbe 5:23–34
60. Nesic D, Miller MC, Quinkert ZT, Stein M, Chait BT, Stebbins CE (2010) Helicobacter

pylori CagA inhibits PAR1-MARK family kinases by mimicking host substrates. Nat Struct Mol Biol 17:130–132

61. Brandt S, Kwok T, Hartig R, Konig W, Backert S (2005) NF-kappaB activation and potentiation of proinflammatory responses by the *Helicobacter pylori* CagA protein. Proc Natl Acad Sci USA 102:9300–9305
62. Kim SY, Lee YC, Kim HK, Blaser MJ (2006) *Helicobacter pylori* CagA transfection of gastric epithelial cells induces interleukin-8. Cell Microbiol 8:97–106
63. Lamb A, Yang XD, Tsang YH, Li JD, Higashi H, Hatakeyama M et al (2009) *Helicobacter pylori* CagA activates NF-kappaB by targeting TAK1 for TRAF6-mediated Lys 63 ubiquitination. EMBO Rep 10:1242–1249
64. Keates S, Keates AC, Warny M, Peek RM, Murray PG, Kelly CP (1999) Differential activation of mitogen-activated protein kinases in AGS gastric epithelial cells by *cag*+ and *cag*- *Helicobacter pylori*. J Immunol 163: 5552–5559
65. Keates S, Sougioultzis S, Keates AC, Zhao D, Peek RM, Shaw LM et al (2001) *Cag*+ *Helicobacter pylori* induce transactivation of the epidermal growth factor receptor in AGS gastric epithelial cells. J Biol Chem 276: 48127–48134
66. Meyer-Ter-Vehn T, Covacci A, Kist M, Pahl HL (2000) *Helicobacter pylori* activates mitogen-activated protein kinase cascades and induces expression of the proto-oncogenes c-fos and c-jun. J Biol Chem 275: 16064–16072
67. Naumann M, Wessler S, Bartsch C, Wieland B, Covacci A, Haas R et al (1999) Activation of activator protein 1 and stress response kinases in epithelial cells colonized by *Helicobacter pylori* encoding the *cag* pathogenicity island. J Biol Chem 274:31655–31662
68. Kaparakis M, Turnbull L, Carneiro L, Firth S, Coleman HA, Parkington HC et al (2010) Bacterial membrane vesicles deliver peptidoglycan to NOD1 in epithelial cells. Cell Microbiol 12:372–385
69. Boughan PK, Argent RH, Body-Malapel M, Park JH, Ewings KE, Bowie AG et al (2006) Nucleotide-binding oligomerization domain-1 and epidermal growth factor receptor: critical regulators of beta-defensins during *Helicobacter pylori* infection. J Biol Chem 281: 11637–11648
70. Viala J, Chaput C, Boneca IG, Cardona A, Girardin SE, Moran AP et al (2004) Nod1 responds to peptidoglycan delivered by the *Helicobacter pylori cag* pathogenicity island. Nat Immunol 5:1166–1174
71. Allison CC, Kufer TA, Kremmer E, Kaparakis M, Ferrero RL (2009) *Helicobacter pylori* induces MAPK phosphorylation and AP-1 activation via a NOD1-dependent mechanism. J Immunol 183:8099–8109
72. Watanabe T, Asano N, Fichtner-Feigl S, Gorelick PL, Tsuji Y, Matsumoto Y et al (2010) NOD1 contributes to mouse host defense against *Helicobacter pylori* via induction of type I IFN and activation of the ISGF3 signaling pathway. J Clin Invest 120:1645–1662
73. Strober W, Murray PJ, Kitani A, Watanabe T (2006) Signalling pathways and molecular interactions of NOD1 and NOD2. Nat Rev Immunol 6:9–20
74. Nagy TA, Frey MR, Yan F, Israel DA, Polk DB, Peek RM (2009) *Helicobacter pylori* regulates cellular migration and apoptosis by activation of phosphatidylinositol 3-kinase signaling. J Infect Dis 199:641–651
75. Wang G, Olczak A, Forsberg LS, Maier RJ (2009) Oxidative stress-induced peptidoglycan deacetylase in *Helicobacter pylori*. J Biol Chem 284:6790–6800
76. Franco AT, Friedman DB, Nagy TA, Romero-Gallo J, Krishna U, Kendall A et al (2009) Delineation of a carcinogenic *Helicobacter pylori* proteome. Mol Cell Proteomics 8:1947–1958
77. Wang G, Maier SE, Lo LF, Maier G, Dosi S, Maier RJ (2010) Peptidoglycan deacetylation in *Helicobacter pylori* contributes to bacterial survival by mitigating host immune responses. Infect Immun 78:4660–4666

Chapter 8

Genetic Manipulation of a Naturally Competent Bacterium, *Helicobacter pylori*

Jennifer M. Noto and Richard M. Peek Jr.

Abstract

Genetic manipulation of *Helicobacter pylori* facilitates characterization and functional analysis of individual *H. pylori* genes. This chapter discusses the methods involved in *H. pylori* chromosomal DNA isolation, mutagenesis of individual genes, and natural transformation.

Key words: *Helicobacter pylori*, Allelic exchange mutagenesis, Natural competence, Transformation

1. Introduction

Helicobacter pylori is a Gram-negative bacterial pathogen that inhabits the human stomach. Infection can lead to a number of diverse disease outcomes, including gastric and duodenal ulceration, gastric adenocarcinoma, and gastric mucosa-associated lymphoid tissue (MALT) lymphoma. It is well documented that both *H. pylori* virulence determinants as well as host factors heighten the risk for *H. pylori*-associated disease. *H. pylori* exhibits remarkable genetic diversity as a species, as evidenced by variation of gene order and genetic content of strains, mosaic nature of genes, and sequence diversity within conserved genes (1–4). *H. pylori* has a panmictic population structure, indicative of frequent genetic exchange among strains. Additionally, there is evidence that recombination between *H. pylori* strains can occur in vivo during naturally occurring mixed infections. Analyses of complete genomic sequences from a number of *H. pylori* isolates indicate that this organism has incorporated genetic material from other organisms during the course of its existence. This exchange of genetic material (e.g., horizontal gene

JeanMarie Houghton (ed.), *Helicobacter Species: Methods and Protocols*, Methods in Molecular Biology, vol. 921,
DOI 10.1007/978-1-62703-005-2_8, © Springer Science+Business Media, LLC 2012

exchange) can occur via three classical mechanisms: conjugation, transduction, and transformation. Conjugation is the transfer of genetic material between bacteria through direct cell-to-cell contact, while transduction involves the transfer of genetic material between bacteria via bacteriophage. Transformation is the genetic alteration of a cell resulting from the uptake, incorporation, and expression of exogenous genetic material. *H. pylori* is, in general, naturally competent, whereby transformation is a relatively frequent occurrence. Approximately 75% of all *H. pylori* strains are naturally competent for uptake of *H. pylori* chromosomal DNA, and this feature allows investigators to easily genetically manipulate this pathogen. Characterization of experimentally generated mutants forms the basis of modern microbial genetics and is particularly useful for determining the function of individual genes by comparing the phenotype of a mutant in which the gene is no longer expressed to the wild-type strain. Targeted or random mutagenesis of *H. pylori* involves a technique termed allelic exchange mutagenesis, in which cloned fragments of *H. pylori* DNA are inactivated in *Escherichia coli* and then introduced into *H. pylori* for allelic exchange by natural transformation. This chapter discusses the specific materials and methods required for creating genetic mutations in the bacterial pathogen, *Helicobacter pylori*.

2. Materials

2.1. H. pylori Chromosomal DNA Isolation

1. Culture of *Helicobacter pylori* (plate or broth culture).
2. TE buffer: 10 mM Tris–HCl, 1 mM EDTA, pH 8.0 (store at 4°C).
3. 10% Sodium dodecyl sulfate (SDS) (store at room temperature).
4. 20 mg/ml Proteinase K (store aliquots at −20°C).
5. 5 M NaCl (store at room temperature).
6. CTAB/NaCl solution (store at room temperature):
   (a) Dissolve 4.1 g NaCl in 80 ml of dH_2O.
   (b) Add 10 g CTAB slowly while heating and stirring.
   (c) Adjust final volume to 100 ml.
7. 24:1 Chloroform/isoamyl alcohol (store at 4°C).
8. 25:24:1 Phenol/chloroform/isoamyl alcohol (store at 4°C).
9. Isopropanol (store at 4°C).
10. 70% Ethanol (store at 4°C).

2.2. H. pylori Mutagenesis

1. *H. pylori* chromosomal DNA.
2. PCR reagents: 10× PCR amplification buffer with $MgCl_2$, 20 mM dNTP mixture, 5 μM primers, *Taq* DNA polymerase, and molecular biology grade H_2O.

3. Commercially available PCR purification kit or, alternatively, reagents for standard method of DNA purification and isolation.
4. PCR cloning vector.
5. Ligation reagents: T4 DNA ligase, ligation buffer, and molecular biology grade H_2O.
6. Restriction enzyme reagents: Restriction enzyme(s), 10× restriction enzyme buffer(s), and molecular biology grade H_2O.
7. Competent *E. coli* cells, Luria broth (LB), and antibiotic stock solutions (see Note 1).
8. Commercially available plasmid preparation kit or, alternatively, reagents for standard alkaline lysis method.
9. Agarose, ethidium bromide, and gel electrophoresis equipment.

2.3. Natural Transformation

1. *Brucella* broth (BB) and *Brucella* agar (BA):
   (a) *Brucella* broth: BB + 10% fetal calf serum (FCS).
   (b) *Brucella* agar plates: BA + 10% FCS.
2. Antibiotic stock solutions (see Note 2).
3. Tryptic soy agar with 5% sheep blood (TSA blood) plates (commercially available).
4. Sterile 1× PBS.
5. Naturally competent strain of *Helicobacter pylori*.

3. Methods

3.1. *H. pylori* Chromosomal DNA Isolation (5)

1. Grow *H. pylori* on TSA blood plates overnight (~16–18 h) at 37°C with 5% CO_2.
2. Suspend *H. pylori* cells from one plate in 900 μl of TE buffer.
3. Pellet bacteria by centrifugation at 8,000 rpm for 5 min.
4. Discard the supernatant and resuspend the bacterial pellet in 567 μl of TE buffer.
5. Add 30 μl of 10% SDS and 3 μl of 20 mg/ml proteinase K to yield a final concentration of 100 μl/ml proteinase K in 0.5% SDS. Mix thoroughly and incubate for 1 h at 37°C.
6. Add 100 μl of 5 M NaCl and mix thoroughly.
7. Add 80 μl of CTAB/NaCl solution. Mix thoroughly and incubate for 10 min at 65°C.
8. Add an equal volume of 24:1 chloroform/isoamyl alcohol (~700–800 μl), mix thoroughly, and subject to centrifugation at 8,000 rpm for 5 min.

9. Transfer aqueous phase (supernatant) to a new microcentrifuge tube, leaving the interface behind.
10. Add an equal volume of 25:24:1 phenol/chloroform/isoamyl alcohol to the aqueous phase, mix thoroughly, and subject to centrifugation at 8,000 rpm for 5 min.
11. Transfer aqueous phase (supernatant) to a new microcentrifuge tube and add 0.6 volume isopropanol to precipitate the nucleic acids. Shake gently until stringy, white DNA precipitate appears.
12. Pellet the DNA by centrifugation at 8,000 rpm for 5 min and then remove isopropanol.
13. Wash the DNA with 100 μl 70% ethanol to remove residual CTAB.
14. Pellet the DNA by centrifugation at 8,000 rpm for 5 min and then remove and discard ethanol.
15. Dry DNA pellet in lyophilizer or, alternatively, open tube and leave on bench or in hood.
16. Once DNA pellet is dry, dissolve DNA in 100 μl TE buffer.

3.2. H. pylori Mutagenesis

3.2.1. Site-Specific Mutagenesis by Overlap Extension (6, 7)

1. Design and synthesize oligonucleotide primers, Forward (F), Reverse Mutagenic (RM), Forward Mutagenic (FM), and Reverse (R), based on the known gene sequence within the *H. pylori* strain of interest (Fig. 1) (see Note 3).
2. In a sterile PCR tube, set up PCR1 by mixing the following reagents:

H. pylori chromosomal DNA	~100 ng
10× amplification buffer	10 μl
20 mM dNTP mixture	1 μl
5 μM F primer (30 pmol)	6 μl
5 μM RM primer (30 pmol)	6 μl
Taq DNA polymerase	1–2 units
Molecular biology grade H_2O	To 100 μl

3. In another sterile PCR tube, set up PCR2 by mixing the following reagents:

H. pylori chromosomal DNA	~100 ng
10× amplification buffer	10 μl
20 mM dNTP mixture	1 μl
5 μM FM primer (30 pmol)	6 μl
5 μM R primer (30 pmol)	6 μl
Taq DNA polymerase	1–2 units
Molecular biology grade H_2O	To 100 μl (see Note 4)

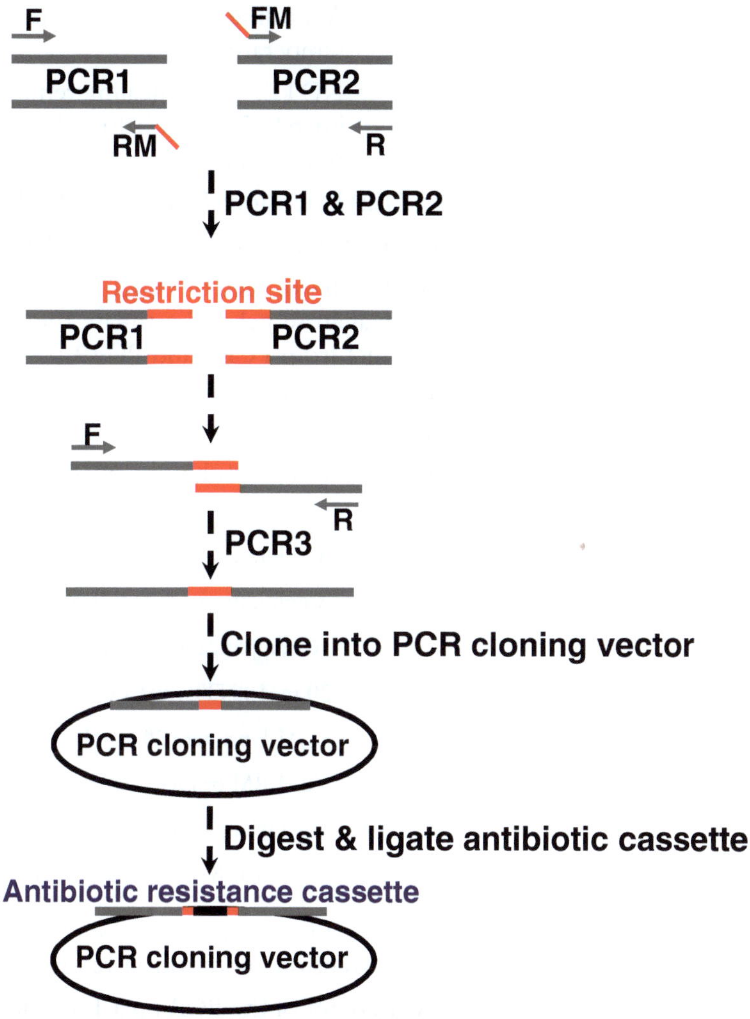

Fig. 1. Schematic representation of site-specific mutagenesis by overlap extension (7). Forward (F) and Reverse Mutagenic (RM) primers are used to amplify the upstream region of a specific target gene in PCR1. Forward Mutagenic (FM) and Reverse (R) primers are used to amplify the downstream region of a specific target gene in PCR2. Mutagenic primers are designed to encode a novel restriction site, as shown. Following the primary PCR steps, products from PCR1 and PCR2 are used in conjunction with Forward (F) and Reverse (R) primers in PCR3, where the overlapping sequences serve to prime this reaction. The product of PCR3 is then cloned into a PCR cloning vector and subsequently digested at the novel restriction sites. An antibiotic resistance-encoding cassette (as shown) is then inserted into this site and the construct is then used for *H. pylori* transformation.

4. Amplify the nucleic acids in PCR1 and PCR2 by using the denaturation, annealing, and extension conditions listed below:
   (a) Initial Denaturation: 5 min at 95°C.
   (b) 30–35 repeated cycles:
      (i) Denaturation: 1 min at 95°C.

(ii) Annealing: 1 min at determined annealing temperature.

NOTE: The annealing temperature should be approximately 5–10°C below the melting temperature (Tm) of primers.

(iii) Extension: 1 min at 72°C (1 min/kb).

(c) Final Extension: 10 min at 72°C.

5. Analyze 5% of each of PCR1 and PCR2 on a 1% ethidium bromide agarose gel and estimate the concentration of amplified product from PCR1 and PCR2.
6. Purify the PCR1 and PCR2 products using a commercially available PCR purification kit.
7. In a sterile PCR tube, set up PCR3 by mixing the following reagents:

Amplification product from PCR1	~50 ng
Amplification product from PCR2	~50 ng
10× amplification buffer	10 µl
20 mM dNTP mixture	1 µl
5 µM F primer (30 pmol)	6 µl
5 µM RM primer (30 pmol)	6 µl
Taq DNA polymerase	1–2 units
Molecular biology grade H_2O	To 100 µl

8. Amplify the nucleic acids using the denaturation, annealing, and extension conditions listed in step 4.
9. Analyze 5% of the PCR on a 1% ethidium bromide agarose gel, estimate the concentration of amplified product from PCR3, and then purify the PCR3 product using commercially available PCR purification kit (see Note 5).
10. Clone PCR3 product into PCR cloning vector and transform competent *E. coli* cells. NOTE: Blue/white screening is a useful tool to identify positive *E. coli* transformants/clones.
11. Grow *E. coli* cultures overnight at 37°C with shaking at ~100 rpm and then harvest plasmid using commercially available plasmid preparation kit.
12. Digest plasmid with appropriate restriction enzyme to open plasmid at engineered, novel restriction site within cloned PCR3 product.
13. Ligate antibiotic resistance-encoding cassette within *H. pylori* gene of interest.
14. Transform competent *E. coli* cells with ligated vector.

15. Grow overnight at 37°C with shaking at ~100 rpm and then select for antibiotic-resistant clones by growing *E. coli* on selective antibiotic LB plates.
16. Harvest *E. coli* and isolate antibiotic-resistant plasmids for *Helicobacter pylori* transformation.
17. To confirm directional cloning, use restriction enzymes to digest PCR cloning vector at various sites within the multiple cloning site.
    (a) Digest with enzymes to excise PCR3 product.
    (b) Digest with enzymes to excise antibiotic resistance-encoding cassette.
    (c) Digest with multiple enzymes to determine directionality of the PCR3 product and antibiotic resistance-encoding cassette.
18. Analyze products of restriction digestions on a 1% ethidium bromide agarose gel and proceed if restriction maps are correct (see Note 6).

3.3. Natural Transformation

3.3.1. Plate Transformation

1. Collect *H. pylori* cells from an overnight (~16–18 h) TSA blood plate in 1 ml of sterile 1× PBS.
2. Harvest bacteria by centrifugation at 8,000 rpm for 5 min.
3. Resuspend bacteria in 100 μl of *Brucella* broth.
4. Spot 25 μl of *H. pylori* cells onto a TSA blood plate.
5. Add ~1 μg of antibiotic-resistant plasmid preparation to *H. pylori*.
6. Incubate TSA plate right-side-up overnight at 37°C with 5% CO_2.
7. Following overnight incubation, use sterile swab to transfer *H. pylori* spot to *Brucella* agar plates with antibiotic for selection of positive transformants.
8. Incubate for ~5 days or until colonies appear.
9. Expand colonies on *Brucella* agar selective antibiotic plates and freeze *H. pylori*-positive transformants at −80°C for future use.
10. Confirm *H. pylori* mutations by PCR amplification of chromosomal DNA and sequence analysis of the gene of interest.

3.3.2. Broth Transformation

1. Inoculate *Brucella* broth with *H. pylori* (either from a 24-h TSA blood plate or from an *H. pylori* starter overnight culture). Cultures should be started at an OD_{600} of ~0.1–0.2 (see Note 7).
2. Add ~1–2 μg of antibiotic-resistant plasmid DNA and grow overnight (~16–18 h) at 37°C with 5% CO_2.
3. Measure OD_{600} (see Note 8).

4. Remove 1 ml of *H. pylori* culture and harvest by centrifugation at 8,000 rpm for 5 min.
5. Resuspend bacteria in 100 μl of *Brucella* broth.
6. Plate 100 μl on *Brucella* agar antibiotic plates and incubate right-side-up overnight at 37°C with 5% CO_2.
7. Turn plate upside down and continue to incubate at 37°C with 5% CO_2 for ~5 days or until colonies appear.
8. Expand colonies on *Brucella* agar selective antibiotic plates and freeze positive transformants at −80°C for future use.
9. Confirm *H. pylori* mutations by PCR amplification of chromosomal DNA and sequence analysis.

4. Notes

1. Antibiotics used for selection of PCR cloning vector and PCR cloning vector with ligated PCR3 product, encoding a novel restriction site and antibiotic resistance cassette.
2. Antibiotic concentrations used for *E. coli* selection may need to be adjusted for *H. pylori* selection.
3. Mutagenic primers should encode a novel restriction site, not present within the gene of interest or in the PCR cloning vector to be used in subsequent steps.
4. Primer concentration may vary depending on the particular type of PCR reagents used. Follow manufacturer's instructions.
5. If more than one band is present, it may be necessary to use a commercially available gel extraction kit to extract and purify appropriate product.
6. Also submit plasmid for sequence analysis.
7. This starting OD_{600} is optimal for obtaining cultures in the log phase of growth after overnight incubation.
8. OD_{600} should be approximately 1.5–2.0.

References

1. Alm RA, Bina J, Andrews BM, Doig P, Hancock RE, Trust TJ (2000) Comparative genomics of *Helicobacter pylori*: analysis of the outer membrane protein families. Infect Immun 68:4155–4168
2. Go MF, Kapur V, Graham DY, Musser JM (1996) Population genetic analysis of *Helicobacter pylori* by multilocus enzyme electrophoresis: extensive allelic diversity and recombinational population structure. J Bacteriol 178: 3934–3938
3. Salama N, Guillemin K, McDaniel TK, Sherlock G, Tompkins L, Falkow S (2000) A whole-genome microarray reveals genetic diversity among *Helicobacter pylori* strains. Proc Natl Acad Sci USA 97:14668–14673

4. Tomb JF, White O, Kerlavage AR, Clayton RA, Sutton GG, Fleischmann RD et al (1997) The complete genome sequence of the gastric pathogen *Helicobacter pylori*. Nature 388:539–547
5. Wilson K (2001) Preparation of genomic DNA from bacteria. Curr Protoc Mol Biol. Chapter 2: Unit 2.4
6. Cormack B (2001) Directed mutagenesis using the polymerase chain reaction. Curr Protoc Mol Biol. Chapter 8: Unit 8.5
7. Horton RM, Cai ZL, Ho SN, Pease LR (1990) Gene splicing by overlap extension: tailor-made genes using the polymerase chain reaction. BioTechniques 8:528–535

Chapter 9

A Method for Short-Term Culture of Human Gastric Epithelial Cells to Study the Effects of *Helicobacter pylori*

Marina Leite and Ceu Figueiredo

Abstract

In vitro studies of *Helicobacter pylori* pathogenesis mostly rely on the use of tumor-derived cell lines. Although invaluable, tumor cell lines are not representative of the normal cell physiology. Thus, the use of primary gastric epithelial cell cultures provides an important tool for investigating the mechanisms underlying *H. pylori* infection, as well as for validating the in vitro findings obtained with tumor-derived cell line models. Here we describe a method for isolation and short-term culture of human primary gastric epithelial cells obtained from gastric biopsy specimens, and the use of these cells to evaluate the effect of *H. pylori* on the junctional adhesion molecule-A protein.

Key words: *Helicobacter pylori*, Primary cell cultures, Human gastric epithelial cells

1. Introduction

Helicobacter pylori infects more than half of the world's population. Although most infected individuals develop chronic gastritis as a consequence of the infection, a proportion of those may develop more severe clinical outcomes such as peptic ulcer disease, gastric mucosa-associated lymphoid tissue lymphoma, and gastric carcinoma (1). Progression towards disease occurs only in some individuals, and seems to be a result of the interactions among bacterial and host factors (2).

The interaction of *H. pylori* with the host gastric epithelium is the basis for the development of disease. In vitro evidence for the effects of *H. pylori* on the epithelium comes mainly from the use of tumor-derived cell lines due to the lack of normal human gastric epithelial cell lines. This constitutes a major limitation, since

JeanMarie Houghton (ed.), *Helicobacter Species: Methods and Protocols*, Methods in Molecular Biology, vol. 921, DOI 10.1007/978-1-62703-005-2_9,

tumor-derived cell lines harbor (epi)genetic alterations in oncogenic and tumor suppressor signaling pathways. One of the most frequent alterations in such cell line models is the disruption of cell–cell adhesion complexes, impairing the formation of cohesive, polarized cell monolayers. Thus, the use of primary gastric epithelial cells is of paramount importance to investigate the mechanisms of *H. pylori* pathogenesis, and to validate findings obtained with tumor-derived cell line models.

Here we describe a method for isolation and culture of human primary gastric epithelial cells obtained from biopsy specimens collected during upper gastroscopy. This method is an adaptation of culture methodologies used to recover epithelial cells and generate primary cultures from the human fetal stomach or from gastric biopsies (3, 4). As illustrated here, this short-term primary culture system can be used for infection experiments with *H. pylori* to verify the influence of the bacteria on tight junction proteins, such as the junctional adhesion molecule-A (JAM-A). Results obtained with this model confirm previous observations showing that *H. pylori* alters the structure of epithelial tight junctions in a canine kidney cell line (5).

2. Materials

2.1. Cell Culture Media and Supplements

1. Hanks' Balanced Salt Solution (HBSS) (1×), liquid, without Calcium and Magnesium.
2. Dulbecco's Modified Eagle Medium: Nutrient Mixture F-12, DMEM/F12 (1:1) medium, 1×, liquid, GlutaMAX™-I.
3. Fetal bovine serum (FBS), Research Grade.
4. Gentamicin (10 mg/mL).
5. Penicillin (10,000 units)–Streptomycin (10,000 μg) solution.
6. Fungizone antimycotic solution.
7. Dithiothreitol (DTT), 1 M.
8. Human epidermal growth factor (EGF). Human EGF is dissolved at 250 μg/mL in deionized sterile filtered water and stored at −20°C in single-use aliquots to avoid repeated freezing and thawing.
9. Human recombinant insulin, zinc solution.

2.2. Reagents and Solutions

1. Collagenase type 1. Dissolve the lyophilized collagenase I in HBSS to obtain a 20,000 U/mL solution. Store in single-use aliquots at −20°C.
2. Neutral protease (Dispase). Dissolve the lyophilized enzyme in HBSS to obtain a 200 U/mL solution. Store in single-use aliquots at −20°C.

3. Trypsin inhibitor (from soybean). Resuspend the enzyme in HBSS to a concentration of 10 mg/mL. Store in single-use aliquots at −20°C.
4. Bovine albumin serum (BSA), Cohn Fraction V. For the cell dissociation solution prepare a 250 mg/mL stock solution of BSA in ultrapure water and store in single-use aliquots at −20°C.
5. Distilled water.
6. Phosphate-buffered saline (PBS), pH 7.4.
7. Cell dissociation solution: HBSS with 200 U/mL of collagenase type 1, 1.2 U/mL of neutral protease (dispase), 0.01 mg/mL of soybean trypsin inhibitor, 1.25 mg/mL of BSA, 0.1 mM DTT. This solution is freshly prepared and sterilized through a 0.2 μm filter membrane.
8. Biopsy collection medium: HBSS supplemented with 0.2% fungizone, 1% penicillin–streptomycin, 10 μg/mL gentamicin.
9. Anti-pancytokeratin fluorescein isothiocyanate (FITC)-conjugated mouse monoclonal antibody (our laboratory uses 1:50 dilution; clone C-11; Sigma-Aldrich Inc., St. Louis, MO, USA).
10. Anti-JAM-A rabbit polyclonal antibody (our laboratory uses 1:30 dilution; Zymed® Laboratories Inc., CA, USA).
11. Goat anti-rabbit-Alexa594 is used as a secondary antibody.
12. Vectashield™ Mounting Medium with DAPI.

2.3. Cell Culture Plastics

1. Lab-Tek® II glass 4-chamber slide system (Nalgene Nunc™ International, IL, USA).
2. T25 cell culture flasks.
3. 15 and 30 mL sterile polypropylene tubes.
4. 0.2 μm cellulose acetate filters.

3. Methods

The establishment of primary gastric epithelial culture is discussed in 3 steps: (1) collection of gastric biopsies; (2) dissociation of gastric biopsy specimens by mechanical and enzymatic procedures; and (3) culture of gastric epithelial cells in medium supplemented with growth factors and hormones. In addition, we describe the protocol for infection of these cells with *H. pylori* and the immunofluorescence for cytokeratins and JAM-A.

3.1. Tissue Collection

All tissue must be collected under an approved protocol (IRB approved) and with full patient consent in accordance with institutional policy.

1. Obtain biopsies of the gastric mucosa from individuals undergoing upper endoscopy, measuring approximately 10 mm^3 each. In this protocol we have used specimens mainly from the gastric antrum.
2. Transfer gastric biopsies (at least 5 specimens) to sterile flat-bottom polypropylene tubes, containing ice-cold biopsy collection medium. Transport biopsies immediately to the laboratory, at 4°C.

3.2. Isolation of Human Gastric Epithelial Cells

All solutions used are sterile or sterilized by filtration and, unless otherwise stated, all procedures are performed inside a laminar flow hood under sterile conditions.

1. Remove the biopsy collection medium in which fragments were transported and rinse the biopsies, 3×, in fresh pre-warmed biopsy collection medium.
2. Place the biopsies into a 10 mm culture Petri dish with 2 mL of pre-warmed biopsy collection medium.
3. Mechanically dissociate gastric biopsies into pieces of approximately 1 mm in size with the help of tweezers and a surgical scalpel, avoiding excessive cutting (see Note 1).
4. Centrifuge the cell suspension at 259 ×*g*, for 5 min, at room temperature, to pellet cells. Discard medium.
5. Add 5 mL of pre-warmed cell dissociation solution, and transfer the cell suspension into a T25 cell culture flask (see Note 2).
6. Place the T25 cell culture flask vertically in a shaking incubator at approximately 150 rpm, at 37°C, for 5–10 min, for the enzymatic dissociation to occur (see Note 3).
7. Transfer the cell suspension into a centrifuge tube and pellet cells by centrifugation (259 ×*g*, for 5 min, at room temperature). Discard the dissociation solution after centrifugation.
8. Wash cells, 2×, in 10 mL of HBSS, and pellet gastric cells by centrifugation (259 ×*g*, for 5 min, at room temperature).

3.3. Short-Term Culture of Gastric Epithelial Cells

1. Resuspend the cell pellet of gastric cells in 2.5 mL of DMEM/F12 (1:1) cell culture medium supplemented with 20% FBS, 0.2% fungizone, 1% penicillin–streptomycin, 10 μg/mL gentamicin, 10 ng/mL EGF, and 4 μg/mL insulin (complete DMEM/F12 medium).
2. Seed 500 μL of the cell suspension into each of the 4 chambers of the Lab-Tek® glass slide (day 0 of culture), and culture in standard culture conditions (37°C under a 5% CO_2 humidified

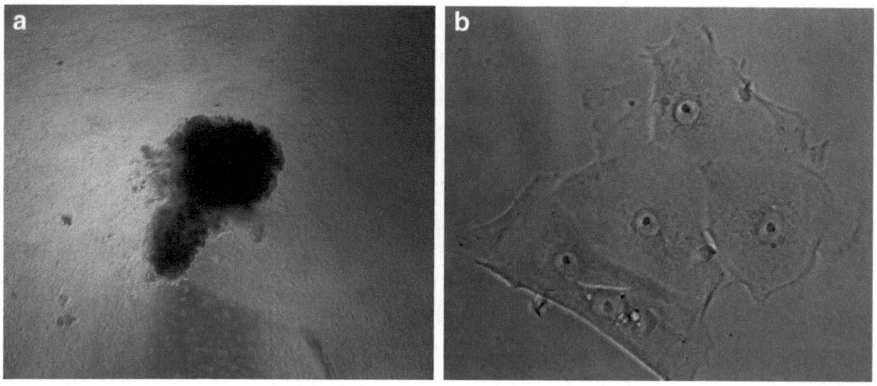

Fig. 1. Brightfield microscopy of human primary gastric cells obtained by mechanical and enzymatic processing of gastric biopsies. (**a**) Cluster of primary gastric cells 24 h after being placed in culture (magnification 50×); (**b**) primary gastric cells in culture for 5 days (magnification 200×).

atmosphere) for 5 days. Let cell suspension sit undisturbed, for at least 24 h, to allow attachment. Figure 1a shows the appearance of cell clusters.

3. On day 1 of culture, remove non-adherent cells by washing 2× each Lab-Tek® chamber with pre-warmed DMEM/F12 (1:1) medium (without supplementation) (see Note 4). Refill each chamber with 500 μL of complete pre-warmed DMEM/F12 (1:1) medium.
4. On day 3 of culture, renew the culture medium, after two washes with pre-warmed DMEM/F12 (1:1) medium (without supplementation), by adding complete pre-warmed DMEM/F12 (1:1) medium with half of the initial concentration of EGF (5 ng/mL) and insulin (2 μg/mL).

3.4. Infection of Primary Gastric Cells with Helicobacter pylori

1. On day 4 of cell culture, before inoculation with *H. pylori*, wash, 3×, the Lab-Tek® chambers with pre-warmed HBSS to remove any non-adherent cells and antibiotics (see Note 5). Figure 1b shows the appearance of gastric epithelial cells after 5 days in culture.
2. Add 500 μL of pre-warmed DMEM/F12 (1:1) plus 10% FBS medium, without further supplementation (antibiotics, antimycotics, growth factors, or hormones).
3. Infect primary gastric cells with *H. pylori* by adding to each of the Lab-Tek® chambers the appropriate volume of bacteria suspension to obtain the desired multiplicity of infection (MOI) (see Note 6). Use the same volume of PBS for control cultures.
4. Let the coculture proceed for 24 h in an incubator at 37°C, under a 5% CO_2 humidified atmosphere.
5. After the coculture period, wash, 3×, the Lab-Tek® chambers with HBSS and fix cells (e.g., ice-cold methanol, for 10 min). Store slides at −20°C until immunofluorescence is performed.

3.5. Immunofluorescence Analysis of Cytokeratins and JAM-A

1. Incubate fixed cells in 5% BSA–PBS for 30 min to block nonspecific binding of antibodies.
2. Remove the blocking solution and incubate cells with the primary antibody mixture, comprising the unconjugated anti-JAM-A rabbit polyclonal and the FITC-conjugated anti-pancytokeratin mouse monoclonal antibodies diluted in 2% BSA–PBS, for 2 h, at room temperature, in the dark (see Note 7).
3. Wash cells in PBS (3×, for 5 min each), with shaking, in the dark.
4. Incubate cells with the fluorescent-conjugated goat anti-rabbit-Alexa594 secondary antibody diluted in 2% BSA–PBS, for 1 h, at room temperature, in the dark. Figure 2 depicts the cytokeratin

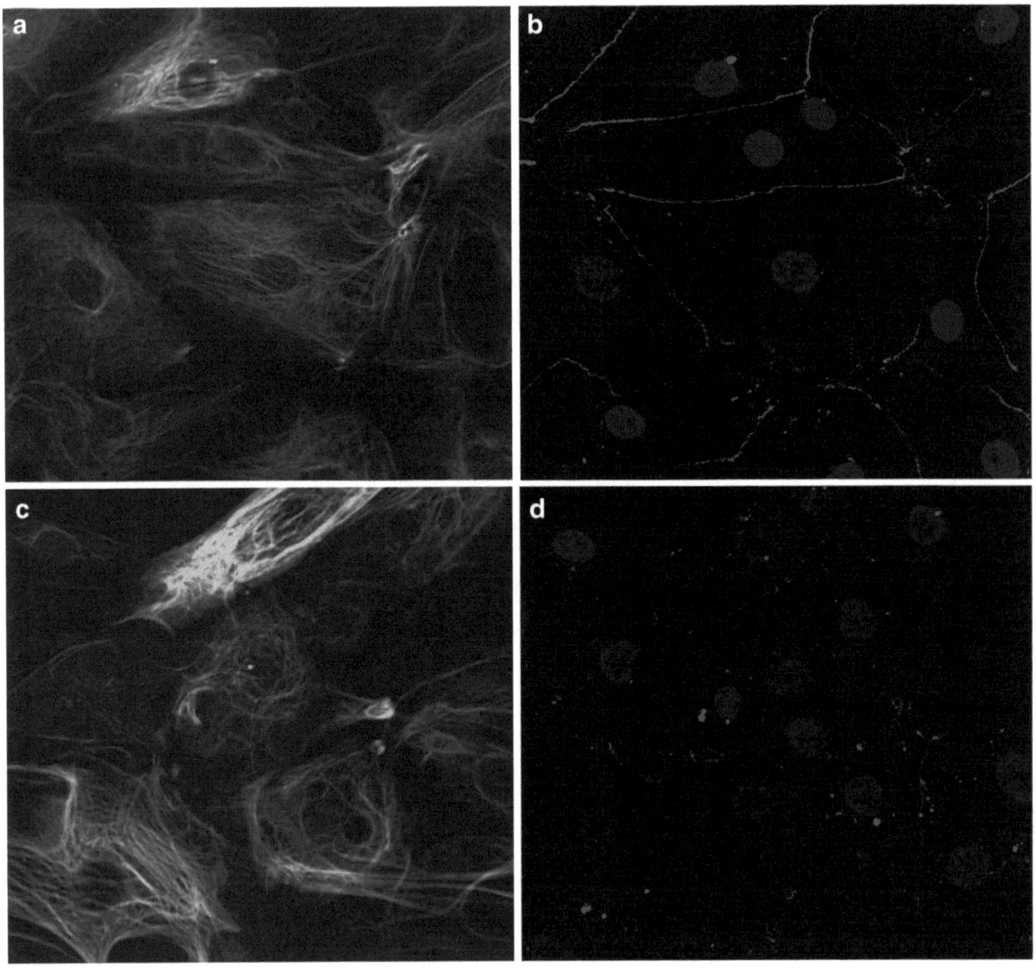

Fig. 2. Immunofluorescence of uninfected (**a** and **b**) and *H. pylori*-infected (**c** and **d**) primary gastric cells (magnification 400×). (**a** and **c**) stained with an anti-pancytokeratin-fluorescein isothiocyanate (FITC)-conjugated antibody; (**b** and **d**) stained with an anti-JAM-A antibody. Nuclei were counterstained with DAPI. All cells are positive for epithelial cytokeratins in the cytoplasm, confirming their epithelial origin. The tight junction protein JAM-A is present at cell–cell contacts in uninfected cells whereas *H. pylori*-infected cells show decreased JAM-A expression at the membrane and delocalization to the cell cytoplasm. Images were captured using a Zeiss Imager Z1 Axiovert S100 microscope and Axiocam camera (Carl Zeiss GmbH, Germany) and represent z-stack projections of 15 to 25 images of 0.25 μm sections.

staining and the immuno-expression of JAM-A in uninfected (A and B) and *H. pylori*-infected (C and D) primary gastric cells.

5. Wash in PBS (3×, for 5 min each), with shaking.
6. Wash once with distilled water, with shaking.
7. Drain slides and mount with Vectashield™ mounting medium with DAPI for nuclear counterstaining. Cover with a wide coverslide and seal to prevent drying and movement under the microscope.
8. Observe slides on a fluorescence microscope.
9. Store slides in the dark at –20°C.

4. Notes

1. During the mechanical dissociation minimize cell damage avoiding extensive dissociation of gastric biopsies into single-cell suspensions. The final density of epithelial cells will depend mostly on the number of clusters initially plated rather than on the proliferation of primary cells. Try also to remove gastric mucus as much as possible during this step.
2. The cell dissociation solution should be prepared fresh to circumvent the decrease of enzymatic activity when stored at 4°C.
3. The time for enzymatic dissociation should be adjusted by microscopic evaluation of cells to ensure that cell clusters are obtained.
4. After 24 h of culture, remove non-adherent cells that are mainly constituted by cells attached to mucus and blood cells. Further washes are important to remove the blood cells that remain in culture.
5. *Helicobacter pylori* strain 26695 (American Type Culture Collection [ATCC] 700392 is cultured in BBL™ Trypticase™ Soy Agar with 5% Sheep Blood (TSA II) (Becton Dickinson France SAS, Le Pont de Claix, France) under a microaerophilic atmosphere, at 37°C, for 48 h. For infection experiments, *H. pylori* is harvested in sterile PBS, pH7.4, and added to primary gastric cultures at the desired MOI; for control cultures, a corresponding volume of sterile PBS is added to cells.
6. Check the confluence of cells in each Lab-Tek® chamber and estimate the number of cells in the chamber area to calculate the concentration of *H. pylori* to be added in order to obtain the desired MOI.
7. In a double immunofluorescence, take into consideration to use primary antibodies generated in different animals (e.g., one antibody generated in mouse and the other generated in

rabbit). Use 300 μL of the primary antibody mix to cover the cells in each Lab-Tek® chamber to avoid evaporation during the incubation period.

Acknowledgments

This study was supported by the ERA-NET Pathogenomics (ERA-PTG/002/2006) and by the Portuguese Foundation for Science and Technology - FCT (PTDC/BIA-MIC/116890/2010). M.L. is supported by a postdoc fellowship (BPD/33420/2008). IPATIMUP is an Associate Laboratory of the Portuguese Ministry of Science, Technology and Higher Education and is partially supported by FCT.

References

1. Atherton JC (2006) The pathogenesis of *Helicobacter pylori*-induced gastro-duodenal diseases. Annu Rev Pathol 1:63–96
2. Polk DB, Peek RM Jr (2010) *Helicobacter pylori*: gastric cancer and beyond. Nat Rev Cancer 10:403–414
3. Chailler P, Ménard D (2005) A new approach to primary culture of human gastric epithelium. Methods Mol Med 107:217–236
4. Smoot DT, Sewchand J, Young K, Desbordes BC, Allen CR, Naab T (2000) A method for establishing primary cultures of human gastric epithelial cells. Methods Cell Sci 22:133–136
5. Amieva MR, Vogelmann R, Covacci A, Tompkins LS, Nelson WJ, Falkow S (2003) Disruption of the epithelial apical-junctional complex by *Helicobacter pylori* CagA. Science 300: 1430–1440

Chapter 10

Cell Culture-Based Assays to Test for Bacterial Adherence and Internalization

Deepa Raju, David Rizzuti, and Nicola L. Jones

Abstract

Adherence and internalization of *Helicobacter pylori* into epithelial cells is a recently recognized event in the pathogen's life cycle.

Key words: Adherence, Internalization, Epithelial cells, Helicobacter

1. Introduction

Helicobacter pylori is a gram-negative microaerophilic organism that infects half of the world's population (1, 2). It is the primary cause of gastritis, peptic ulcers, as well as gastric cancer. Recent studies have shown that *H. pylori* is able adhere to and internalize within gastric epithelial and immune cells and cause persistent infection (3, 4). For example, studies have demonstrated the presence of *H. pylori* within gastric epithelial cells by electron microscopy analysis of gastric biopsy samples obtained from infected humans (3, 5–10). The ability of *H. pylori* to invade mammalian epithelial cells has also been demonstrated in gastric adenocarcinoma-derived cell lines. A study by Oh et al. identified intracellular bacteria in gastric epithelial progenitor cells in a murine model of infection (11). Studies by Amieva and colleagues have shown the formation of large vacuoles post *H. pylori* invasion of epithelial cells in which the bacteria can persist for long periods of time (3). These studies were confirmed in our lab where we observed that persistence of *H. pylori* was dependent on the VacA toxin and that the large vacuoles acted as intracellular niches for the organism to survive and persist (4). Furthermore, *H. pylori* infection triggers a predominantly T-Helper type 1 immune

JeanMarie Houghton (ed.), *Helicobacter Species: Methods and Protocols*, Methods in Molecular Biology, vol. 921,
DOI 10.1007/978-1-62703-005-2_10, © Springer Science+Business Media, LLC 2012

response that is characteristic of intracellular pathogens (12). These studies collectively illustrate the significance of invasion during *H. pylori* infection exemplifying the importance of understanding the methods required to investigate this host–pathogen interaction.

The aim of the following two chapters is to outline the methodology for studying the adherence and internalization of bacteria in model epithelial cells. The first chapter describes techniques in cell biology that are used to study and enumerate the number of organisms that can invade and gain entry into a model epithelium. The following chapter then looks at multiple assays that can be performed to assess the effect of bacterial invasion and bacterial effector molecules such as CagA and VacA on epithelial cell signaling pathways.

2. Materials

Bacterial strains and model epithelial cells.

2.1. Bacterial Strains of Helicobacter pylori

A common strain of *Helicobacter pylori* used in culture and infection studies is *H. pylori* strain 60190 (provided by Dr. Peek, Vanderbilt, TN). This is a lab-adapted strain of *H. pylori* that can produce both the virulence factors VacA and CagA ($VacA^+$, $CagA^+$). Isogenic *vacA* mutant (strain 60190, $VacA^-$, $CagA^+$) and isogenic CagA mutant (60190, $VacA^+$, $CagA^-$) strains are also used to elucidate the effect of these virulence factors on bacterial adhesion, infection, and internalization.

2.2. Epithelial Cells

The epithelial cells chosen for the examination of the direct effects of bacteria are based on (1) similarity to the epithelium in the target host organ that the bacteria invades and (2) the type of assay to be performed. In our lab, we have made extensive use of the gastric epithelial cell line AGS (ATCC#CRL-1739) originally derived from human gastric adenocarcinoma, to study the effects of the bacteria (4, 13, 14). HEp-2 laryngeal epithelial cells can be used when signaling pathways of interest are aberrant in gastric cancer cell lines. As an example, the effect of CagA on the STAT3 pathway cannot be investigated in gastric cancer cell lines as STAT3 is constitutively activated (13). For studies investigating autophagy, Murine embryonic fibroblasts (MEFs) were also employed (14).

3. Culture Reagents

3.1. Bacterial Culture and Storage

Materials necessary for the growth of *H. pylori* strains:

1. Columbia blood agar plates with 5% sheep blood (Oxoid Biotech, MP0351, Nepean, ON).

2. Bacterial inoculating loops (10 μL, VWR, 12000-810).
3. Heat-inactivated fetal bovine serum (FBS) (Invitrogen).
4. Brucella broth (Brucella broth base dissolved in double-distilled H_2O, Sigma Life Science) supplemented with 10% FBS.
5. Erlenmeyer culture flasks (125 mL, BD Falcon).
6. Sterile glycerol.
7. Sterile cryovials (2 mL, Simport, T310-2A, Beloeil, QC).
8. Brucella agar plates (1% casein, 1.5% peptic digest, 0.1% dextrose, 0.2% yeast extract, 0.5% sodium chloride, 0.01% sodium bisulfite, 1.5% agar).
9. Kanamycin (Invitrogen, 15160-054).
10. Ham's F12 medium (Wisent, 318-010-CL).
11. 1.5 mL Microtubes (Diamed, Mississauga, ON, Lot # 8509).
12. Cell scraper (Sarstedt, Newton, NC, Lot # 9293400).

3.2. Epithelial Cell Culture

Human gastric epithelial cells (AGS) are thawed from stock and cultured in Ham's F-12 media (Wisent Technologies, PQ, Canada) supplemented with 10% FBS and incubated at 37°C in a CO_2 environment. MEFs and HEp-2 cells are grown in Dulbecco's modified eagle medium (D-MEM) containing FBS (10%). Trypsin–EDTA (0.05%) (Wisent Technologies, PQ, Canada) is used to detach cells from plasticware such as T-75 and T-25 flasks and sterile phosphate-buffered saline (PBS) is used for washing.

4. Equipment

Major pieces of equipment include:

1. 4°C Refrigerator.
2. 70°C Freezer.
3. Centrifuge.
4. Level 2 biosafety cabinet.
5. 37°C Incubator.
6. 37°C Shaker/incubator.
7. Light microscope.
8. Microaerophilic incubator or evacuation anaerobic jars.
9. Other plasticware required for cell culture include T-75 flasks, T-25 flasks, 6-well and 12-well plates, coverslips, etc.

5. Methods

5.1. Bacterial Culture

Helicobacter pylori are cultured on Columbia blood agar plates with 5% sheep blood for 72 h, under microaerophilic conditions, at 37°C. Frozen bacterial stocks are inoculated (via sterile plastic inoculating loop) on blood agar plates and then placed in a microaerophilic incubator.[1] The isogenic mutant strains are cultured on Brucella agar plates supplemented with 10% FBS and 20 μg/ml kanamycin. For experiments that require infection, the cultured *H. pylori* strains are harvested from the agar plates with a sterile inoculating loop and resuspended in Brucella broth (Sigma Life Science) supplemented with 10% FBS. The bacteria are grown for 24 h at 37°C in an Erlenmeyer culture flask under microaerophilic conditions with shaking at 120 rpm. Prior to infection, bacteria should be examined under a light microscope to check for motility, health, and contamination. The bacteria are centrifuged at 3,800 × *g* for 10 min to form a pellet, which is then resuspended in Ham's F12 medium with 10% FBS to an optical density of 0.5.

5.2. Bacterial Infection of Epithelial Cells

Human gastric epithelial cells are cultured in Ham's F12 medium with 10% FBS at 37°C in a CO_2 incubator and then placed in 6-well tissue culture plates to a confluency of 80–90%. *H. pylori* (cultured as described above) is then added to the plates at a multiplicity of infection (MOI) of 100:1 and allowed to adhere to the epithelial cells and invade for 4 h. The non-internalized bacteria are then washed off with sterile PBS. The epithelial cells are then incubated in Ham's F12 medium supplemented with 100 μg/ml gentamicin (Wisent Technologies, 450-135-XL) for 1 h. The gentamicin kills any remaining non-internalized bacteria. After incubation, the media supplemented with gentamicin is removed, cells are washed with PBS, and the cells are incubated in Ham's F12 medium supplemented with 10 μg/ml gentamicin to prevent any extracellular bacterial growth. The infected cells are incubated at 37°C in a CO_2 incubator for 24 h.

6. Enumeration of Adherent and Invading Bacteria

6.1. Counting Bacteria by Plating

The most traditional method to assess if bacteria and in particular *H. pylori* is able to multiply within gastric epithelial cells is by performing a gentamicin assay as described above followed by rupturing the

[1] If a microaerophilic incubator is not available, evacuation anaerobic jars can be used to attain the necessary microaerophilic conditions by flushing a microaerophilic gaseous mixture into the sealed jars following evacuation. The sealed jars are then placed in a 37°C incubator for 72 h. Long-term storage of *H. pylori* can be achieved by adding 10% sterile glycerol to the Brucella bacterial culture broth, aliquoting the mixture into sterile 2 mL cryovials (Simport, T310-2A, Beloeil, QC), and storing at −70°C.

epithelial cells and plating the media to allow for the internalized bacteria to multiply and count the number of colonies formed. The detailed protocol for enumeration by plating is as follows:

1. *H. pylori* strains and epithelial cells are cultured as described in Subheadings 3.1 and 3.2.
2. Bacteria are quantitated by measuring optical density and then added at the desired MOI to epithelial cells as described in Subheading 5.2. In our lab, the MOI used is 100:1.
3. At the end of the 24 h invasion in the presence of gentamicin to kill any external bacteria, epithelial cells are washed with PBS and incubated for 5–10 min at 37 C in 1% saponin. Saponin is a detergent that under short exposures ruptures epithelial cells to release internalized bacteria. To avoid killing of the bacteria by saponin, the process needs to be completed quickly to avoid exposure to saponin for prolonged periods.
4. The saponin solution is then serially diluted and 100 μl of the desired dilution is plated on Brucella agar plates supplemented with 10% FBS.
5. Brucella agar plates are then incubated in microaerophilic chambers at 37 C for 48–72 h.
6. A single bacterium on the plate multiplies to become a single colony; hence the number of bacterial colonies multiplied with the dilution factor will determine the number of bacteria that invade the epithelial cells.

6.2. Enumeration Using Electron Microscopy

One of the disadvantages of plating is that it is a general representative of both adherent and invasive bacteria. Therefore, the subcellular localization of bacteria within the cell cannot be determined. Direct visualization using either electron microscopy or confocal microscopy is employed to study the cellular location of invading bacteria and to determine the compartments within which the bacteria can multiply. The technique most commonly employed for visualization of invading bacteria and localization within cellular compartments is electron microscopy. Immunoelectron microscopy may also be used in cases where a particular compartment containing the bacteria needs to be identified. For this chapter, the following protocol focuses on the preparation of samples for electron microscopy and their visualization. Representative cells in the grid can then be counted for both adherent and invading bacteria.

Preparation of samples

1. AGS cells are cultured in 10 cm culture dishes overnight at 37°C and grown to confluency in Ham's F-12 medium as described in the above sections. Required strains of *H. pylori* are grown overnight in Brucella broth.

2. Epithelial cells are infected with *H. pylori* at an MOI of 100:1. Infected cells are incubated for 24 h at 37°C. Uninfected cells are used as controls.
3. After 24 h, cells are washed twice with PBS to remove external bacteria and fixed using glutaraldehyde.
4. Fixed cells are then cut into thin sections and processed for electron microscopy. Images are acquired on a transmission electron microscope (Electron Microscope Facility, Hospital for Sick Children) at a magnification of 4,000–15,000 times.
5. Bacteria are counted in representative cells. Adherent bacteria can be differentiated from internalized bacteria and counted separately using this technique.

6.3. Immuno-fluorescence

Another technique by which bacteria can be enumerated is by immunofluorescence.

External adherent bacteria can be differentiated from internalized bacteria using a two-step staining process where external bacteria are labeled with one tag before permeabilizing the epithelial cells followed by the immunolabeling of both internal and adherent bacteria with a different tag post permeabilization. The difference between the total number of stained bacteria and the bacteria that stained before permeablization (i.e., adherent bacteria) is equivalent to the number of internal bacteria. The general protocol for immunolabeling to identify internal versus adherent bacteria is as follows:

1. AGS cells are grown on coverslips in 12-well plates in Ham's F12 medium with 10% FBS and bacteria grown overnight as described earlier.
2. The next day cells are infected with *H. pylori* and a gentamicin invasion assay as outlined in Subheading 5.2 performed for different time points.
3. After incubation with the bacteria for the required time, cells are washed with PBS and fixed with 4% paraformaldehyde (diluted 1 in 4 from a 16% solution) (Sigma Aldrich, St. Louis, MO).
4. Cells are then blocked with 5% milk solution in PBS for 1 h after which they are incubated with a rabbit anti-*H. pylori* antibody (Dako cytomation, Denmark) for 1 h at a dilution of 1:100.
5. The cells are then washed in PBS and incubated with a Cy5 conjugated anti-rabbit secondary antibody (dilution 1:1,000) (Jackson Immunoresearch laboratories, West Grove, PA).
6. Cells are washed in PBS and then permeabilized using 0.1% triton. The staining procedure described above is repeated using a Cy3-conjugated anti-rabbit secondary antibody.
7. After washing in PBS, the coverslips are then mounted onto microscope slides using fluorescent mounting medium (Dako Cytomation, Ontario, Canada).

8. Cells are then viewed using a Spinning Disk Confocal Microscope with a Leica DMIRE2 fluorescence microscope, a Hamamatsu Back-Thinned EM-CCD camera, spinning disk confocal scan, 4 laser lines, an ASI motorized stage, and an Improvision Piezo Focus Drive. The equipment is controlled by Volocity acquisition software (Improvision) on an Apple Power Mac G5. Confocal images are analyzed and bacteria can be counted using the Volocity software program (Improvision) on a Macintosh computer.

7. Conclusions

Until very recently, *H. pylori* had been considered a noninvasive pathogen. However, recent studies have shown the ability of *H. pylori* to both adhere to and invade gastric epithelial cells in vitro and in vivo (3, 4, 15, 16). These studies however are still in their infancy and more detailed studies are now essential to fully understand the mechanisms involved in entry and survival of the pathogen in a model gastric epithelium as well as the biologic significance in vivo. Conclusions from these studies will lead to a more complete understanding of the bacterial factors and pathways that contribute to *H. pylori* disease pathogenesis.

Acknowledgments

NLJ is supported by a Canadian Institute for Health Research operating grant (No. 17886).

References

1. Peek RM Jr, Blaser MJ (2002) Helicobacter pylori and gastrointestinal tract adenocarcinomas. Nat Rev Cancer 2:28–37
2. Peek RM Jr, Crabtree JE (2006) Helicobacter infection and gastric neoplasia. J Pathol 208:233–248
3. Amieva MR, Salama NR, Tompkins LS, Falkow S (2002) Helicobacter pylori enter and survive within multivesicular vacuoles of epithelial cells. Cell Microbiol 4:677–690
4. Terebiznik MR et al (2006) Helicobacter pylori VacA toxin promotes bacterial intracellular survival in gastric epithelial cells. Infect Immun 74:6599–6614
5. Dubois A (1995) Spiral bacteria in the human stomach: the gastric helicobacters. Emerg Infect Dis 1:79–85
6. Wyle FA, Tarnawski A, Schulman D, Dabros W (1990) Evidence for gastric mucosal cell invasion by C. pylori: an ultrastructural study. J Clin Gastroenterol 12(Suppl 1): S92–98
7. Petersen AM, Blom J, Andersen LP, Krogfelt KA (2000) Role of strain type, AGS cells and fetal calf serum in Helicobacter pylori adhesion and invasion assays. FEMS Immunol Med Microbiol 29:59–67
8. Petersen AM, Krogfelt KA (2003) Helicobacter pylori: an invading microorganism? A review. FEMS Immunol Med Microbiol 36: 117–126
9. Petersen AM, Sorensen K, Blom J, Krogfelt KA (2001) Reduced intracellular survival of Helicobacter pylori vacA mutants in comparison

with their wild-types indicates the role of VacA in pathogenesis. FEMS Immunol Med Microbiol 30:103–108
10. Rittig MG et al (2003) Helicobacter pylori-induced homotypic phagosome fusion in human monocytes is independent of the bacterial vacA and cag status. Cell Microbiol 5:887–899
11. Oh JD, Karam SM, Gordon JI (2005) Intracellular Helicobacter pylori in gastric epithelial progenitors. Proc Natl Acad Sci USA 102:5186–5191
12. Rieder G, Fischer W, Haas R (2005) Interaction of Helicobacter pylori with host cells: function of secreted and translocated molecules. Curr Opin Microbiol 8:67–73
13. Bronte-Tinkew DM et al (2009) Helicobacter pylori cytotoxin-associated gene A activates the signal transducer and activator of transcription 3 pathway in vitro and in vivo. Cancer Res 69:632–639
14. Terebiznik MR et al (2009) Effect of Helicobacter pylori's vacuolating cytotoxin on the autophagy pathway in gastric epithelial cells. Autophagy 5:370–379
15. Bjorkholm B et al (2000) Helicobacter pylori entry into human gastric epithelial cells: a potential determinant of virulence, persistence, and treatment failures. Helicobacter 5:148–154
16. Correa P, Houghton J (2007) Carcinogenesis of Helicobacter pylori. Gastroenterology 133:659–672

Chapter 11

Cell Culture Assays to Evaluate Bacterial Toxicity and Virulence

Deepa Raju, David Rizzuti, and Nicola L. Jones

Abstract

Helicobacter pylori CagA and VacA are two critical virulence factors that modulate disease severity in the infected host. The following chapter outlines methods employed to study the effects of these virulence factors on several key signaling pathways in epithelial cells.

Key words: CagA, VacA, STAT3, Autophagy

1. Introduction

Helicobacter pylori produces a variety of virulence factors; two of the most investigated include the cytotoxin-associated gene A (CagA) and the vacuolating cytotoxin (VacA) (1). CagA is a bacterial effector protein that is injected into the host cell via the Type IV secretion system. The presence of infection with a CagA+strain is associated with an increased risk of gastric cancer (1–3). VacA is an 88 kDa pore-forming toxin that is secreted by the bacteria (4). The presence of infection with the toxin is associated with intracellular survival and persistent infection with *H. pylori* (5) as well as increased disease severity. Investigation of the signaling pathways that are altered by these virulence factors has shed light on mechanisms that contribute to disease pathogenesis during chronic infection. Here, we outline a variety of assays that can be employed to study the effects of these virulence factors on epithelial cells. We focus on specific examples of signaling pathways that we have shown to be affected such as the STAT3 signaling pathway or the autophagy pathway, which are modulated by CagA and VacA, respectively (6, 7).

JeanMarie Houghton (ed.), *Helicobacter Species: Methods and Protocols*, Methods in Molecular Biology, vol. 921,
DOI 10.1007/978-1-62703-005-2_11,

2. Materials and Culture Reagents

Materials necessary for the growth of *H. pylori* strains:

1. Columbia blood agar plates with 5% sheep blood (Oxoid Biotech, MP0351, Nepean, ON).
2. Bacterial inoculating loops (10 μl, VWR, 12000-810).
3. Heat-inactivated fetal bovine serum (FBS) (Invitrogen).
4. Brucella broth (Brucella broth base dissolved in double-distilled H2O, Sigma Life Science) supplemented with 10% FBS.
5. Erlenmeyer culture flasks (125 ml, BD Falcon).
6. Sterile glycerol.
7. Sterile cryovials (2 ml, Simport, T310-2A, Beloeil, QC).
8. Brucella agar plates (1% casein, 1.5% peptic digest, 0.1% dextrose, 0.2% yeast extract, 0.5% sodium chloride, 0.01% sodium bisulfite, 1.5% agar).
9. Kanamycin (Invitrogen, 15160-054).
10. 1.5 ml Microtubes (Diamed, Mississauga, ON, Lot # 8509).
11. Cell scraper (Sarstedt, Newton, NC, Lot # 9293400).
12. For preparation of VacA toxin culture supernatants, protein concentration tubes are essential (30 kDa filter, Millipore Corp, USA).
13. Cell culture reagents including growth media such as Dulbecco's minimal essential medium and Ham's F-12 (Wisent, Canada), SDS-PAGE gel running apparatus, and Western blotting apparatus (Biorad, USA).
14. Other miscellaneous items required are 6- and 12-well plates for cell culture, T-75 flasks, T-25 flasks, glass coverslips, and other plastic and glassware.

3. Studying the Effects of CagA on Epithelial Cells

CagA is the virulence factor that is most strongly associated with increased risk for gastric cancer. The *cagA* gene that encodes CagA is present within the cag pathogenicity island (cag PAI), a 40 kb DNA segment that encodes for approximately 32 genes including the components of the type IV secretion system (8). Upon infection of gastric epithelial cells, CagA is translocated into host cells via the type IV secretion system and localizes to the inner leaflet of the plasma membrane. At the plasma membrane, CagA can undergo tyrosine phosphorylation at the EPIYA motif leading to a cascade of signaling events including activation of the Erk-MAP kinase pathways which contribute to cell transformation and induction of gastric cancer phenotypes (9, 10). CagA also affects cell signaling independent of tyrosine phosphorylation by interacting with cellular junction proteins such as ZO-1 (11). CagA also activates NF-kB leading to a mitogenic signaling response (12). Recent studies in our laboratory have demonstrated that CagA activates STAT3, a key signaling molecule that may promote the progression of gastric cancer during infection with *H. pylori* (6).

These examples illustrate the intense interest in understanding the mechanisms by which CagA can promote tumorigenesis. Researchers have utilized both in vitro and in vivo models to study these cellular pathways. Below is a description of some of the common epithelial cell culture-based assays that can be used to study the effects of cagA on host cells. The effects of cagA on the STAT3 pathway have been used as an example to illustrate these techniques.

3.1. Effects on Cell Signaling Proteins Using Western Blotting

Western blotting can be used to determine the effects of *H. pylori* infection on internal cell signaling in epithelial cells. For example, this technique can be used to determine whether *H. pylori* activates the Signal Transducer and Activator of Transcription 3 (STAT3) pathway in epithelial cells (6).

1. After 24 h of infection with *H. pylori* (as described in Chapter 10), epithelial cells are harvested using a cell scraper (Sarstedt, Newton, NC, Lot # 9293400) and P1000 micropipette, and transferred into corresponding 1.5 ml microtubes (Diamed, Mississauga, ON, Lot # 8509).
2. The cells are isolated via centrifugation and the supernatants are collected into newly labeled microtubes and placed in a –70°C freezer to examine cytokine production at a later time. The cells are then resuspended in 200 μl PBS to wash off any remaining media.
3. Centrifugation is repeated and the cells resuspended in 100 μl RIPA buffer (NaF, NaCl, PMSF, RIPA, Na_3VO_4) to lyse the epithelial cell membranes.
4. After 20 min, the lysates are isolated by centrifugation, and the supernatants are collected into a new corresponding eppendorf, as the supernatants contain the intracellular proteins and signaling molecules.
5. The supernatants (50 μl) are then loaded onto a 10% SDS-Polyacrylamide gel along with 10% lamelli buffer, run via electrophoresis, and transferred to a nitrocellulose membrane.

6. The nitrocellulose membrane is then incubated in 5% milk (made with Tris-Buffered Saline Solution supplemented with 0.05% Tween—TBST) at room temperature for 30 min, to block any nonspecific antibody binding. The membrane can be cut to allow immunoblotting of both actin (43 kDa) and phosphorylated STAT3 (PSTAT3, 88 kDa) bands.
7. The actin portion is incubated in milk supplemented with a primary antibody specific for actin (at a concentration of 1:5000) overnight, at 4°C. Similarly the PSTAT3 portion of the band is incubated in milk supplemented with a primary antibody specific for PSTAT3 (at a concentration of 1:200) overnight, at 4°C.
8. The membranes are then washed 3 times for 5 min in TBST.
9. The membranes are then incubated in milk supplemented with secondary antibody specific for the primary antibodies (at a concentration of 1:5,000) for 1 h at room temperature.
10. The membranes are then washed 3 times for 20 min in TBST followed by incubation with a Western Blotting Luminol Reagent sc-2048 kit (Santa Cruz Biotechnology, Inc., Santa Cruz, CA) for 3 min, and the membranes are exposed (using Amersham Hyperfilm ECL—High performance chemiluminescence film) to examine protein levels.

3.2. Effects on Cell Signaling Using Transient Transfection and Immuno fluorescent Microscopy

To study the localization of CagA within epithelial cells, transient transfection of CagA tagged to fluorophores such as green fluorescent protein can be employed. Additionally, immunofluorescence can be used to determine the effects of CagA on signaling pathways within the cell. For example, an antibody specific to phosphorylated STAT3 can be employed to observe the phosphorylation and translocation of STAT3 into the nucleus of the host cell post infection with *H. pylori* or transfection with CagA-GFP (6). Detailed protocols for transfection of epithelial cells and immunofluorescence are described below.

3.2.1. DNA Transfections

1. Epithelial cells are grown to 40–50% confluency in Dulbecco's minimum essential medium supplemented with 10% FBS.
2. Cells are transfected with 1 μg of CagA-GFP DNA using transfection agents such as Fugene HD (3 μg) (Roche, USA) or using the AMAXA nucleofector System according to the procedure outlined (AMAXA Biosystems, MD, USA).
3. Cells are incubated with transfection mixture and medium for at least 20 h to allow optimal CagA expression with limited cytotoxity.

3.2.2. Immunofluorescence

The general protocol for immunolabelling is as follows:

1. AGS cells are grown on cover slips in 12-well plates in Ham's F12 medium with 10% FBS. Bacteria are grown overnight as described earlier. Cells are then transfected with CagA-GFP or infected with *H. pylori* as required.
2. After incubation for the required time, cells are washed with PBS and fixed with 4% paraformaldehyde (diluted 1 in 4 from a 16% solution) (Sigma Aldrich, St. Louis, MO).
3. Nonspecific labeling in cells is then blocked with 5% milk solution in PBS for 1 h followed by incubation with a rabbit anti-phospho-STAT3 or STAT3 antibody (1:250 dilution, Cell signaling, MA, USA) for another 1 h.
4. The cells are then washed in PBS and incubated with a Cy3- or Cy5-conjugated secondary antibody (dilution 1:1,000) (Jackson Immunoresearch laboratories, West Grove, PA).
5. After washing in PBS, the cover slips are then mounted onto microscope slides using fluorescent mounting medium (Dako Cytomation, Ontario, Canada).
6. Cells are then viewed using a Spinning Disk Confocal Microscope with a Leica DMIRE2 fluorescence microscope, a Hamamatsu Back-Thinned EM-CCD camera, spinning disk confocal scan, 4 laser lines, an ASI motorized stage, and an Improvision Piezo Focus Drive. The equipment is controlled by Volocity acquisition software (Improvision). Confocal images are analyzed and bacteria can be counted using the Volocity software program (Improvision).

4. Studying the Effects of VacA on Epithelial Cells

VacA is a pore-forming toxin (4, 13) secreted by *H. pylori*. The action of VacA on host cells is characterized by the formation of large vacuoles, hence the name vacuolating toxin (4, 13). These vacuoles occupy a large portion of the cell and form an intracellular niche for the survival of the bacterium. *H. pylori* secretes VacA as monomers which then oligomerize at the cell membrane to form membrane channels. Secreted VacA is then endocytosed via glycosylphosphatidylinositol anchored proteins enriched endosomal compartments (GEECs) from which it traffics into late endosomal and lysosomal compartments and induces vacuolation through a mechanism dependent on Rab7, a small GTPase (14, 15). Recent studies in our lab demonstrate that VacA also affects the autophagy pathway (7). VacA also has been shown to affect the degradative capacity of lysosomes by targeting lysosomal hydrolases such as cathepsin-D (5, 16). The following sections outline the general protocols used to prepare the VacA toxin and study its effects on epithelial cells, focusing on assays to determine vacuolation and induction of autophagy in host cells.

4.1. Preparation of Culture Supernatants

VacA can be concentrated from bacterial cell-free supernatants of *H. pylori* growth medium or can be purified for use. It should be noted that the toxin is inactive in its purified form requiring acid activation prior to use. Bacterial media supernatants can be concentrated as follows to yield a solution with a high concentration of VacA.

1. To prepare culture supernatants, wild-type and mutant *H. pylori* are grown in Brucella broth (Difco Laboratories, Detroit, MI) supplemented with 10% FBS for 24 h at 37°C in a microaerophilic environment in a shaking incubator at 120 rpm.
2. Bacterial cultures are then centrifuged at 3,800 × *g* for 20 min to separate the bacterial pellet from the culture supernatants.
3. The culture supernatants are filtered through a 0.22 μm filter and concentrated 10 times in Ham's F12 media supplemented with 10% FBS using a 30 kDa cutoff Amicon Ultra centrifugal filter (Millipore). The culture supernatants are diluted a further 5 times in Ham's F12 prior to addition to cells for intoxication assays (5, 7).

4.2. Test for Vacuolation Using Neutral Red Assay

The extent of vacuolation in epithelial cells and the effects of various pharmacological agents on vacuolation can be assessed using the uptake of the acidophilic dye, neutral red (7). To perform this assay,

1. AGS cells are grown in 6- or 12-well plates to 100% confluency as described previously.
2. Cells are then treated with the toxin for the required amount of time.
3. A 0.5% stock solution of neutral red dye (Sigma-Aldrich, MO, USA) is prepared in PBS. The stock is further diluted ten times (1:10) before each experiment in cell culture medium containing 10% FBS.
4. The media is removed from the cells and replaced with 100 μl of the staining solution per well for 5 min.
5. Cells are then washed twice with PBS and the neutral red dye extracted from cells using 100 μl of acidified alcohol per well.
6. The optical density at 540 nm is determined using a plate reader. All assays should be performed in triplicate and the mean OD calculated using the OD of the medium alone as background.

4.3. Assessing Vacuolation Through Bright Field Microscopy

Another technique used to determine vacuolation and the size of vacuoles is bright field microscopy. Cells are grown on cover slips and treated with VacA culture supernatants as described in Subheading 4.1. Cells are then mounted onto space ships and observed under bright field microscopy. The size of the vacuoles can be assessed by using the measurement tool of the Volocity program.

4.4. Effects of VacA on the Autophagy Pathway

One of the key pathways in the cell that is affected by VacA is the autophagy pathway. VacA induces autophagy in a manner dependent on its pore-forming ability. Upon acute exposure of cells to VacA toxin (4–6 h), autophagy is induced and actively modulates the levels of toxin within the cells. Thus during acute exposure the autophagy pathway eliminates the toxin leading to a loss of cell vacuolation (7). It should, however, be noted that autophagosomes are distinct from the larger vacuoles formed by VacA. A variety of assays can be employed to investigate the induction of autophagy.

The most commonly used marker for autophagy is the GFP-tagged microtubule-associated protein 1 light chain 3 (LC3) protein. Native LC3 or LC3-I is cytoplasmic. Upon the induction of autophagy, LC3 is lipidated by phosphotidyl ethanolamine (LC3-II), which is then recruited to the autophagosomal membrane by the Atg protein complex of Atg5, 12, and 16. Due to its specificity for autophagosomes, both endogenous LC3 and LC3 tagged to various fluorophores such as GFP and RFP have been extensively used to study autophagy.

4.4.1. Transmission Electron Microscopy to Observe Presence of Early- and Late-Stage Autophagosomes

Electron microscopy allows direct visualization of autophagosomes and allows observation of the cargo within autophagosomes. Samples are prepared as described in Chapter 10, Subheading 6.2. For enumeration of autophagosomes and to distinguish between early autophagosomes and late autophagosomes, the following protocol can be employed:

1. Autophagosomes observed within the cells are classified as early (Ai) or late autophagosomes (Ad) based on the structure of the autophagosome as described in previous studies (17). Early autophagosomes are identified by the presence of intact organelles such as mitochondria within a multilamellar compartment in the cell while double membrane structures consisting of dense degrading organelles are identified as late autophagosomes.
2. Samples of cells are prepared as described in Chapter 10, Subheading 6.2 and mounted on electron microscopy grids. Grids are then randomly sampled by first taking pictures at a lower magnification (4,000×) and then zooming in on the cell

at a higher magnification (12,000–15,000×). Each cell is saved as a picture file for further analysis of autophagosomes.

3. The number of autophagosomes in a single cell is counted for each image and classified as early or late autophagosomes.
4. The total cell area is calculated using the Volocity software program and the number of autophagosomes per cell calculated using the equation outlined in the following reference (17).
5. In cells infected with wild-type *H. pylori* or treated with supernatants from VacA + *H. pylori*, the large single membrane vacuoles induced by the VacA toxin and the characteristic multilamellar autophagosomes can be distinguished.
6. Statistical analysis is performed using a Student's *t*-test on the program GraphPad Prism 4 for Macintosh computers.

4.4.2. Detection of LC3-GFP Puncta by Immunofluorescence

As described previously, LC3-GFP is used extensively as a marker for autophagosomes. LC3-GFP is transfected into AGS cells or MEFs using Fugene HD similar to the transfection of CagA-GFP described in Subheading 3.2.1. Following treatment with VacA, cells are fixed using 4% Formaldehyde solution in PBS and mounted onto microscopic slides. Cells treated with rapamycin or nutrient starvation can serve as positive controls for autophagy. Cells are then observed using a Spinning disk confocal microscope. In untreated cells, LC3-GFP is present in the soluble form, hence is present diffusely in the cytoplasm. Upon induction of autophagy LC3-GFP is observed as distinct puncta distributed throughout the cytoplasm which reflect autophagosomes.

4.4.3. Western Blotting to Observe Conversion of LC3-I to LC3-II Upon Induction of Autophagy

LC3 conversion following autophagy can also be detected via Western blotting. LC3 in its native form is a 19 kDa soluble protein, while in the LC3II form it is a modified 16 kDa protein. The differences in the molecular weight between the two forms can be used to detect the induction of autophagy. Lysates are collected from cells that have been treated with the VacA toxin and run on an SDS-PAGE gel as described in Subheading 3.1. Induction of autophagy is detected by the conversion of LC3I to LC3II.

5. Additional Assays

5.1. Assays for Effect of Bacterial Infection or Bacterial Toxins on Cytokine Production

As described above in Subheading 3.1, cytokine production in epithelial cell supernatants from infected and control cells can be assessed by an Enzyme-Linked Immunosorbent Assay (ELISA). The ELISA can be employed to assess the production of a variety of specific cytokines, such as IL-6. A standard human IL-6 ELISA kit (Invitrogen, Catalog # KCH0061) can be used to quantify the level of human IL-6 in the cell culture medium. The general protocol for ELISA is as follows:

1. First 100 μl of standards, controls, and samples are added to the 96-well tissue culture plate included in the kit. The wells in this 96-well plate are coated with an antibody specific for human IL-6. The IL-6 present in the supernatants will adhere to the anti-IL-6 antibodies coating the bottom of the well.
2. Next 50 μl of a biotinylated anti-IL-6 antibody is added to the wells, and the plate is incubated for 2 h at room temperature. The biotinylated anti-IL-6 antibody will adhere to the IL-6 already adherent to the anti-IL-6 antibody coating the bottom of the well.
3. The wells are then aspirated and washed 4× with the wash solution included in the kit.
4. The wells are then incubated in 100 μl Streptavidin-HRP working solution for 30 min at room temperature. The streptavidin-HRP (streptavidin conjugated to horseradish peroxidise) is a homotetramer that forms high affinity non-covalent bonds with the biotin on the biotinylated anti-IL-6 antibodies.
5. The solution in the wells is then aspirated and the wells washed 4× before a final 30 min incubation with 100 μl of stabilized Chromogen at room temperature. The chromogen forms a colored compound upon oxidation, which can be viewed with a 450 nm wavelength of light.
6. After the 30 min incubation 100 μl of Stop Solution is added to each well and the plate is read at 450 nm. The levels of IL-6 in each well are compared with the standards to determine the total amount of human IL-6 produced by the epithelial cells in response to *H. pylori* infection. This technique can be used to examine the production of other cytokines, such as IL-4, IL-12, or TGF-α, in cell culture.

6. Assays for Effects of Bacterial Toxins on Cell Viability and Survival

Bacterial virulence factors can also affect cell survival. In general cell death can occur by necrosis or apoptosis induced by a variety of stimuli. Apoptotic cell death is characterized morphologically by blebbing of the plasmalemma without loss of integrity of the cell membrane along with condensation of the chromatin and fragmentation of the nucleus. Eventually the cell breaks up into apoptotic bodies. In contrast during necrosis the cell membrane loses its integrity and the cell undergoes swelling whilst maintaining its overall shape. A variety of techniques can be employed both to assess cell viability as well as to specifically assess apoptotic cell death, which are described below.

6.1. MTT Live/Dead Cell Assay

The MTT assay is a quick assay that can be used to determine cell viability under a variety of conditions including treatment with bacterial toxins or during infection of epithelial cells.

1. Cells are grown to confluency and infected or treated with toxins or pharmacological agents for the required time period.
2. An in vitro toxicology MTT assay (Sigma-Aldrich, MO, USA) is used to assess cell death. Cells are washed and treated with 3-(4-5-dimethylthiazol-2-yl)-2,5-diphenyl tetrazolium bromide (MTT) for 1–2 h at 1:10 dilution in culture medium. Viable cells in the well contain mitochondrial dehydrogenase, which cleaves the tetrazolium substrate ring, yielding purple formazan crystals.
3. The degree of enzymatic activity is determined by dissolution of the formazan precipitate using spectrophotometric analysis (OD 540 nm) of the resulting solution. Cells with media alone are used as background reading.

6.2. Analysis of Cell Death by Flow Cytometry

Propidium iodide (PI) is a fluorescent agent that is normally excluded from viable cells as it is membrane impermeant. However, when cells undergo necrosis, the plasma membrane becomes more permeable allowing the PI to enter and intercalate with DNA fluorescing red in the process. Annexin V labels external phosphotidyl serine, which undergoes translocation to the outer plasma membrane upon induction of apoptosis. Co-staining of annexin V and PI can be used to distinguish apoptotic cells (annexin V positive only) and necrotic cells (PI positive). Fluorescent labeling can be detected using flow cytometry (FACS).

1. For FACS, cells are grown in 6 or 12 well plates to confluency and treated with toxins as described above.
2. Post treatment, cells are trypsined using 0.25% trypsin-EDTA solution and transferred into a FACS tube.
3. Cells are then washed twice with 1% bovine serum albumin solution (BSA) in PBS followed by staining with propidium iodide (1:1,000) (Sigma Aldrich, MO, USA) and annexin V (1:1,000) (Molecular Probes, Invitrogen, USA) for 1/2 h.
4. Cells are washed two more times after staining and resuspended in 300 μl of 1% BSA and the staining intensity determined by flow cytometry on a FACS calibur machine (BD Biosciences). The percentage of stained cells is determined and compared with appropriate negative controls.

6.3. Analysis of Apoptosis by TUNEL Staining

1. Epithelial cells are rinsed in sterile cold (4°C) PBS, trypsinized (+EDTA) from the culture plates and enumerated. Then 1×10^5 to 5×10^5 cells are cyto-spun on aptex-coated microscope slides and air-dried (30 min at room temperature). Preparations are rehydrated and endogenous peroxidase is blocked by a 30 min incubation in 0.01% (v/v) H_2O_2.
2. Sections are then transferred to buffer containing proteinase K (20 mg/ml; Boehringer Mannheim) for 15 min, followed by 3×2 min washes in distilled water and then incubated in terminal transferase buffer (200 mM potassium cacodylate, 25 mM Tris–HCl (pH 6.6), 0.2 mM EDTA and 0.25 mg/ml BSA for 5 min at room temperature.
3. The preparations are then incubated at 37°C for 60 min in transferase buffer with the addition of cobalt chloride (1 mM), biotin 16-dUTP (0.01 nM), and terminal transferase (0.5 U/μl; Boehringer Mannheim).
4. Subsequently, the reaction is stopped by transfer to a 300 mM sodium chloride + 30 mM sodium citrate solution and positivity visualized via indirect immunocytochemistry employing avidin-conjugated peroxidase and reaction with diaminobenzidine. Preparations are counterstained in hematoxylin, mounted in Permount and apoptotic cells (i.e., positive brown stain) identified micoscopically and enumerated on a per monolayer basis, or in the case of cytospins, per 100 cells observed.

7. Conclusions

In this chapter, we have attempted to describe a variety of approaches to study the effect of *H. pylori* bacterial virulence factors on epithelial cells. Significant inroads are being made into the development of newer techniques and protocols for studying the effects of the bacteria and its various virulence factors on the host cell. All these approaches together will lead to a more significant understanding of the molecular mechanisms involved in *H. pylori* pathogenesis.

Acknowledgments

NLJ is supported by a Canadian Institute for Health Research operating grant (No. 17886).

References

1. Peek RM Jr, Crabtree JE (2006) Helicobacter infection and gastric neoplasia. J Pathol 208:233–248
2. Peek RM Jr (2004) Clandestine intracellular delivery of Helicobacter pylori CagA: guess who's coming to dinner? Gastroenterology 127:669–672
3. Peek RM Jr, Blaser MJ (2002) Helicobacter pylori and gastrointestinal tract adenocarcinomas. Nat Rev Cancer 2:28–37
4. Cover TL, Blanke SR (2005) Helicobacter pylori VacA, a paradigm for toxin multifunctionality. Nat Rev Microbiol 3:320–332
5. Terebiznik MR et al (2006) Helicobacter pylori VacA toxin promotes bacterial intracellular survival in gastric epithelial cells. Infect Immun 74:6599–6614
6. Bronte-Tinkew DM et al (2009) Helicobacter pylori cytotoxin-associated gene A activates the signal transducer and activator of transcription 3 pathway in vitro and in vivo. Cancer Res 69:632–639
7. Terebiznik MR et al (2009) Effect of Helicobacter pylori's vacuolating cytotoxin on the autophagy pathway in gastric epithelial cells. Autophagy 5:370–379
8. Covacci A, Rappuoli R (2000) Tyrosine-phosphorylated bacterial proteins: Trojan horses for the host cell. J Exp Med 191:587–592
9. Higashi H et al (2002) Biological activity of the Helicobacter pylori virulence factor CagA is determined by variation in the tyrosine phosphorylation sites. Proc Natl Acad Sci USA 99:14428–14433
10. Higashi H et al (2004) Helicobacter pylori CagA induces Ras-independent morphogenetic response through SHP-2 recruitment and activation. J Biol Chem 279:17205–17216
11. Amieva MR et al (2003) Disruption of the epithelial apical-junctional complex by Helicobacter pylori CagA. Science 300: 1430–1434
12. Brandt S, Kwok T, Hartig R, Konig W, Backert S (2005) NF-kappaB activation and potentiation of proinflammatory responses by the Helicobacter pylori CagA protein. Proc Natl Acad Sci USA 102:9300–9305
13. Cover TL, Krishna US, Israel DA, Peek RM Jr (2003) Induction of gastric epithelial cell apoptosis by Helicobacter pylori vacuolating cytotoxin. Cancer Res 63:951–957
14. Gauthier NC et al (2007) Early endosomes associated with dynamic F-actin structures are required for late trafficking of H. pylori VacA toxin. J Cell Biol 177:343–354
15. Papini E et al (1997) The small GTP binding protein rab7 is essential for cellular vacuolation induced by Helicobacter pylori cytotoxin. EMBO J 16:15–24
16. Satin B et al (1997) Effect of helicobacter pylori vacuolating toxin on maturation and extracellular release of procathepsin D and on epidermal growth factor degradation. J Biol Chem 272:25022–25028
17. Eskelinen EL (2008) Fine structure of the autophagosome. Methods Mol Biol 445: 11–28

Chapter 12

Rodent Models of *Helicobacter* Infection, Inflammation, and Disease

Songhua Zhang and Steven F. Moss

Abstract

Animal models are essential for in vivo analysis of Helicobacter-related diseases. Transgenic mice and Mongolian gerbil models have been the corner stone of present research focusing on both bacterial virulence factors and host response to infection. Establishing a reproducible rodent model of persistent *Helicobacter pylori* infection that resembles the *H. pylori*-associated gastritis observed in humans was a considerable challenge until Lee et al. (Gastroenterology 112:1386–1397, 1997) successfully adapted a clinical Cag A- and Vac A-expressing strain for the mouse stomach. This so-called SS1 (Sydney) strain has since been extensively used for *H. pylori* research; other rodent-adapted *Helicobacter* strains have subsequently been developed and utilized in wild-type and genetically engineered rodent models. These bacteria include both *H. pylori* and the larger but related species *H. felis* (originally isolated from cats). In this chapter we focus mainly on these two *Helicobacter* strains and review the rodent models that have been employed to investigate how *Helicobacter* species induce gastric inflammation and disease.

Key words: Mice, Transgenic mice, Gerbil

1. Mice

Most investigators have chosen to use mouse models because of their widespread availability (including multiple inbred strains and genetically engineered variants), short breeding cycles, and the accessibility of experimental reagents. The importance of the host response in determining disease outcome is evident from the diverse range of inflammatory and epithelial responses to experimental *Helicobacter* infections in different murine strains (2–4).

It is important to note that the morphology of the gastric neoplasia in mice is similar to but not identical to that in humans. To ensure uniformity in reporting criteria, investigators are encouraged

JeanMarie Houghton (ed.), *Helicobacter Species: Methods and Protocols*, Methods in Molecular Biology, vol. 921,
DOI 10.1007/978-1-62703-005-2_12, © Springer Science+Business Media, LLC 2012

to follow standard definitions of gastrointestinal malignancy in rodents (5) and to use established scoring systems to enumerate differences between experimental groups (6).

1.1. Wild-Type Mice

1.1.1. C57BL/6 Mice

This strain has been particularly extensively studied for investigating *Helicobacter pylori's* role in gastric carcinogenesis, due in part to the many genetically engineered knockouts available on this background. Following *H. felis* or *H. pylori* infection, the immune response is predominantly Th1-skewed, with a relatively high level of epithelial cell damage and compensatory hyperproliferative response, associated with low bacterial loads (1, 7). *H. felis* induces more severe gastric inflammation in C57BL/6 mice than is observed with *H. pylori* (4).

Whereas in humans the gastritis caused by *H. pylori* inflammation is characterized by the accumulation of neutrophils and mononuclear cells in the mucosa (8), neutrophil recruitment is much less prominent in *H. felis* or *H. pylori*-infected mice. Both *H. pylori*-infected C57BL/6 mice and BALB/c mice show a marked influx of mononuclear cells (3, 4).

H. felis infection can eventually progress via metaplasia and dysplasia to cancer in C57BL/6 mice (9), thus mimicking the morphological sequence of changes observed during gastric carcinogenesis in humans (10). The development of high grade dysplastic lesions was not observed in C57BL/6 mice infected with *H. pylori* SS1, even up to 80 weeks post-infection, the longest term study reported to date (11).

The origins of the neoplastic cells in C57BL/6 mice infected by *H felis* appears to be from bone marrow-derived cells recruited to the site of chronic gastric inflammation induced by *H. felis.* Compelling experimental data point to such bone marrow-derived populations repopulating the gastric niche normally occupied by gastric epithelial progenitors and contributing directly to gastric cancer development (12).

1.1.2. BALB/c Mice

BALB/c mice are widely used in both cancer and immunology studies. In contrast to the C57BL/6 strain, BALB/c mice exhibit a Th2-predominant response to *Helicobacter* infection, characterized by higher bacterial colonization levels but fewer epithelial lesions (2, 3). Lymphocytic aggregation is a characteristic feature in BALB/c mice after long-term *H. pylori* infection (3), thus these mice have been used as a model of *Helicobacter*-induced MALT lymphoma. Following 22 months *H. felis* infection, 38% of BALB/c mice developed this neoplasm, which was not observed in uninfected controls (13). As in humans, antibiotic therapy can eradicate both the infection and the associated lymphoma, demonstrating that chronic *Helicobacter* infection can stimulate antigen-dependent formation of B cell MALT lymphoma in susceptible hosts (14). MALT lymphomas can also be induced by infection with certain isolates of *H. pylori* in Balb/c mice, though not with SS1 (15).

1.1.3. C3H Mice

C3H/HeJ mice carry a mutation in their toll-like receptor 4 gene, rendering them insensitive to lipopolysaccharide. They are relatively easy to colonize with *H. pylori*, including with the fully sequenced strain 26695 or by isogenic mutants deficient in Lewis X or Y expression (16).

Antral colonization with *H. pylori* SS1 was found to be only moderate in C3H/HeJ mice and, compared with infection in C57BL/6 mice, induced relatively little gastric body atrophy after 6 months of infection (4). In contrast, others have reported relatively high bacterial colonization levels in C3H/HeJ mice infection with the same *H. pylori* strain SS1 (17).

1.2. Knockout or Transgenic Mice

Many types of genetically engineered mouse models have been used to gain experimental insights into the immunopathogenesis of *Helicobacter* infection and to develop models of *H. pylori*-associated gastric cancer. Most have been employed on the C57BL/6 background. Some of the more informative models that typify the approaches used to dissect the pathogenesis of the response to *H. pylori* are described below.

1.2.1. INS-GAS Mice

Helicobacter infection in humans is accompanied by mild hypergastrinemia. INS-GAS mice have been engineered on the inbred FVB/N strain to overexpress the human gastrin gene under control of an insulin promoter, resulting in sustained hypergastrinemia. These mice spontaneously develop gastric atrophy and eventually gastric adenocarcinoma, a process that can be rapidly accelerated by experimental infection with *Helicobacter* species (18). This has proven to be a valuable model to investigate the possible synergistic effects of hypergastrinemia and *Helicobacter* infection in gastric carcinogenesis.

Interestingly, male INS-GAS mice have a much more rapid and significant inflammatory and neoplastic response to *H. pylori* infection, to a high-salt diet (7.5%), and the combination of both diet and infection compared with female INS-GAS mice (18–20).

There is now considerable evidence that concurrent extragastric inflammation in mice can impact *Helicobacter*-induced gastric mucosal damage and also other pathological changes in the stomach (21–23). However, whereas intestinal *Helminth* infection tends to reduce inflammatory response to *Helicobacter* species (21), Houghton et al. have reported that coexisting infection of *Toxoplasma gondii* and *H. felis* in BALB/c mice altered the specific *H. felis* immune response and increased interferon gamma (IFN-γ), Interleukin-12 (IL-12) with reduced Interleukin-10 (IL-10) expression, leading to a more severe gastric inflammatory response (23). Other gastric species may also modulate the inflammatory response to *H. pylori*. For example, compared to a pure monoculture of *H. pylori* in male germ-free INS-GAS mice, *H. pylori*-infected INS-GAS mice with complex gastric microbiota led to more severe gastritis and accelerated intraepithelial neoplasia (24).

1.2.2. IFN-γ and Tumor Necrosis Factor Alpha Knockout Mice

The severity of gastritis and epithelial changes are significantly reduced following *H. felis* or *H. pylori* infection in IFN-γ deficient mice compared with wild-type C57BL/6 mice (25, 26), demonstrating the critical role of IFN-γ in *Helicobacter*-induced gastric inflammation and epithelial cell damage. Studies in tumor necrosis factor alpha (TNF-α) knockout mice have given less clear-cut results: one group reporting TNF-α to be required for *Helicobacter felis*-induced gastritis (25) while another found that although *H. pylori* colonization was increased in both IFN-γ- and in TNF-α-deficient mice, the gastric inflammatory response was not decreased in the absence of TNF-α (27).

1.2.3. Interleukin-1 Beta Transgenic Mice

Since El-Omar et al. (28) first demonstrated a link between host genetic factors affecting Interleukin-1 Beta (IL-1β) function and the risk of cancer following *H. pylori* infection (28), there has been considerable interest in exploring the role of this acid-inhibiting, pro-inflammatory cytokine in gastric atrophy and adenocarcinoma development. Synergism between *Helicobacter* infection and IL-1β expression in the promotion of gastric neoplasia was demonstrated by Tu et al. (29) in studies of aging *Helicobacter felis*-infected C57BL/6 mice transgenic for IL-1β overexpression in gastric parietal cells.

1.2.4. IL-10 Knockout Mice

As an anti-inflammatory and immunoregulatory cytokine, IL-10 is an important regulator of the mucosal immune response in vivo. IL-10 deficient 129/EvSv mice infected with *H. felis* developed severe hyperplastic gastritis with epithelial cell hyperproliferation and dedifferentiation within 4 weeks of infection. This was associated with a *Helicobacter*-specific Th1 immune response (30). Thus inhibitory mechanisms including IL-10, probably serve to regulate over-exuberant immune and inflammatory responses to *Helicobacter* species in vivo.

1.2.5. Fas Antigen Transgenic Mice

Epithelial cell apoptosis is an important component of the gastric mucosal response to *H. pylori* infection. To investigate the role of the Fas antigen signaling pathway on *Helicobacter*-infected gastric mucosal growth alterations, Houghton et al. (31) infected Fas antigen-deficient (lpr) mice on a C57BL/6 background with *H. felis.* Although the Fas knockouts exhibited similar inflammatory responses to wild-type C57BL/6 mice, Fas-deficient mice did not undergo mucosal cell apoptosis or gastric atrophy. However, when Fas-deficient mice were reconstituted with wild-type bone marrow cells (to obviate early death in the lpr model), gastric cancers were frequently observed after several months of *H. felis* infection (32). This suggests a critical role for Fas-induced signaling in the gastric mucosa in the prevention of a neoplastic response to a chronic *Helicobacter* infection.

1.2.6. p27-Deficient Mice

Loss of the cyclin-dependent kinase inhibitor $p27^{kip1}$ is a common finding in many human cancers and is associated with poor prognosis, including in gastric cancer (33). *H. pylori* infection is associated with decreased expression of p27 in human gastric epithelial cells (34), and mice lacking the p27 tumor suppressor protein are tumor-prone when exposed to environmental carcinogens (35). After infection with *H. pylori* SS1, p27-deficient mice develop metaplasia, dysplasia, and then gastric cancer after 60 weeks (36).

1.2.7. Cag A-Transgenic Mice

Murine models of *Helicobacter* infection are problematic for elucidating the function of the putative *H.pylori* oncogenes CagA since *H. felis* lacks the cag pathogenicity island encoding many genes important for the function of *H. pylori's* type IV secretory system. Although these genes are present in the genome of SS1, they are functionally deficient for CagA translocation. To circumvent this problem CagA transgenic mice have been engineered to express CagA ubiquitously or predominantly in the stomach resulting in hematological and gastrointestinal malignancies, albeit at low frequency (37).

2. Mongolian Gerbils

Mongolian gerbils have been used to study the effects of *H. pylori* infection by several groups worldwide. Severe gastritis, intestinal metaplasia, gastric ulcers and gastric carcinoma have all been reported in *H. pylori*-infected gerbils (38–40), though some investigators have not been able to reproduce these findings, most likely because the rodents are outbred and the colonies used for experimental manipulations therefore not genetically identical.

As early as 3 weeks after infection with *H. pylori*, inflammatory cells are recruited to the gastric mucosa; consequently there is foveolar hyperplasia, parietal cell loss, and the development of mucous cells expressing trefoil factor 2. Inflammation is more prominent in the gastric antrum than in the fundus (41). This distribution of inflammation in gerbils is similar to that observed early in humans, but contrasts with the corpus-predominant bacterial colonization and inflammation reported in either *H. felis* or *H. pylori* infected C57BL/6, BALB/c, and C3H mice (4).

The first report of any gastric malignancy developing in experimentally infected animals was published in 1998 by Watanabe and colleagues (42). After 62 weeks of infection with *H. pylori* strain TN2GF4, 10 of 27 (37%) of the infected Mongolian gerbils developed gastric carcinoma with histological similarity to human intestinal type gastric cancer. In the same year Honda et al. reported similar findings in a small cohort of gerbils infected by the *H. pylori* type strain NCTC 11637 (ATCC 43504) (43). However, other investigators who could not reproduce these findings have questioned whether some of these apparently malignant neoplasms may represent instead heterotopic proliferative glands (44).

Peek et al. infected outbred Mongolian gerbils with wild-type, CagA-deleted, or Vac A-deleted isogenic mutants of *H. pylori* strain G1.1. Increased apoptosis in the gastric antrum was evident early in infection (2–4 weeks post infection) and *H. pylori*-induced inflammation was accompanied by altered gastric epithelial cell cycling and antral epithelial growth, which was associated with elevated serum gastrin levels (45). Using a *H. pylori* whole genome microarray, the intact cag pathogenicity island was implicated as being critical in the differential host inflammatory responses of the gastric ulcer-associated B128 strain and the duodenal ulcer strain G1.1 (46).

While Mongolian gerbils have provided a practical model for studying the pathogenesis of *Helicobacter* infection, especially gastric cancer, the lack of transgenic or specific gene knockout strains and limited commercially available immunological reagents have limited the utilization of these animals in comparison with mice.

3. Chemicals as Gastric Cocarcinogens with *H. pylori* Infection

Gastric carcinoma is a multistep and multifactorial disease (47). Epidemiological evidence points to a role for dietary components, particularly salt and nitrate intake in conjunction with *H. pylori* as increasing the risk for gastric cancer development (47, 48). Therefore many groups have coadministered certain chemical carcinogens to enhance or accelerate the effects of experimental *Helicobacter* infections in studies of gastric carcinogenesis in rodent models.

Synergy between N-methyl-N-Nitrosourea (MNU) or N-methyl-N′-nitro-N-nitrosoguanidine (MNNG) and *H. pylori* has been shown to increase gastric tumor incidence in Mongolian gerbils (49, 50) and in mice (51, 52).

4. Use of Rodent Models to Develop Gastric Cancer Prevention Strategies

The strong link between *H. pylori* infection and the subsequent development of distal gastric cancer has spurred many intervention studies in humans. Meta-analysis of the diverse and generally underpowered individual trials reported to date indicate that *H. pylori* eradication will probably reduce gastric cancer incidence by about 50%, especially if given relatively early in disease progression

(53). Problems with recruitment and the enormous expense of undertaking these large and lengthy clinical studies have made insights from murine models attractive for the investigation of interventions to reduce gastric cancer incidence. Questions regarding the optimal timing of eradication therapy, the advantage of combined eradication/chemoprevention strategies, and the utility of therapeutic or preventive targeted *H. pylori* vaccines as an anticancer strategy all lend themselves to investigation in rodent models.

For example, *H. pylori* eradication has been shown to markedly reduce stomach cancer incidence in C57BL/6 mice, hypergastrinemic INS-GAS mice and in Mongolian gerbils (8, 54, 55). Majority of gastric MALT lymphoma can similarly be cured by antibiotics at an early stage in mice (13), as it can be in most humans (56).

In *H. pylori* infected, cancer-prone, male hypergastrinemic INS-GAS mice a recent study has shown that a combination of the cyclo-oxygenase 2 selective anti-inflammatory drug sulindac at a dose of 400 ppm in the drinking water and an antimicrobial antibiotic eradication regimen significantly decreased proinflammatory cytokine expression in the stomach and the development of gastric carcinoma (57).

Rodents may also be helpful in evaluating *H. pylori* vaccines as cancer-preventing strategies. For example, Delyria et al. recently achieved a high immunization rate in C57BL/6 mice and granulocyte colony–stimulating factor (G-CSF) knockout C57BL/6 mice by intranasal administration of *H. pylori* SS1 lysate with cholera toxin as an adjuvant. Strong IFN-γ expression and enhanced Th17 producing T-cell responses were observed in the immunized mice, which may be important for neutrophil recruitment and subsequent phagocytosis of *H. pylori* bacteria (58).

5. Conclusions

The development of robust murine and Mongolian gerbil models to investigate the effects of *Helicobacter* infection has allowed investigators to examine the importance of host factors, environmental factors, and bacterial strain virulence in the outcome of gastric diseases. This has greatly enhanced our knowledge of the pathogenesis of chronic gastritis and gastric cancer. These rodent models are also being explored to develop novel therapeutic strategies against *H. pylori* infection and thereby limit the related diseases.

Consideration of the host, the *Helicobacter* strain and environmental microbial and chemical cofactors are all important for optimal translation of these findings to the clinic.

References

1. Lee A, O'Rourke J, De Ungria MC et al (1997) A standardized mouse model of *Helicobacter pylori* infection: introducing the Sydney strain. Gastroenterology 112:1386–1397, PMID: 9098027
2. Mohammadi M, Redline R, Nedrud J, Czinn S (1996) Role of the host in pathogenesis of *Helicobacter*-associated gastritis: *H. felis* infection of inbred and congenic mouse strains. Infect Immun 64:238–245, PMID: 8557346
3. Thompson LJ, Danon SJ, Wilson JE et al (2004) Chronic *Helicobacter pylori* infection with Sydney strain 1 and a newly identified mouse-adapted strain (Sydney strain 2000) in C57BL/6 and BALB/c mice. Infect Immun 72:4668–4679, PMID: 15271928
4. Sakagami T, Dixon M, O'Rourke J et al (1996) Atrophic gastric changes in both *Helicobacter felis* and *Helicobacter pylori* infected mice are host dependent and separate from antral gastritis. Gut 39:639–648, PMID: 9026476
5. Boivin GP, Washington K, Yang K et al (2003) Pathology of mouse models of intestinal cancer: consensus report and recommendations. Gastroenterology 124:762–777, PMID: 12612914
6. Rogers AB, Houghton J (2009) *Helicobacter*-based mouse models of digestive system carcinogenesis. Methods Mol Biol 511:267–295, PMID: 19347301
7. Wang TC, Goldenring JR, Dangler C et al (1998) Mice lacking secretory phospholipase A2 show altered apoptosis and differentiation with *Helicobacter felis* infection. Gastroenterology 114:675–689, PMID: 9516388
8. Dixon MF, Genta RM, Yardley JH, Correa P (1996) Classification and grading of gastritis. The updated Sydney System. International Workshop on the Histopathology of Gastritis, Houston 1994. Am J Surg Pathol 20:1161–1181, PMID: 8827022
9. Cai X, Carlson J, Stoicov C et al (2005) *Helicobacter felis* eradication restores normal architecture and inhibits gastric cancer progression in C57BL/6 mice. Gastroenterology 128:1937–1952, PMID: 15940628
10. Correa P, Haenszel W, Cuello C, Tannenbaum S, Archer M (1975) A model for gastric cancer epidemiology. Lancet 2:58–60, PMID: 49653
11. Kim DH, Kim SW, Song YJ et al (2003) Long-term evaluation of mice model infected with *Helicobacter pylori*: focus on gastric pathology including gastric cancer. Aliment Pharmacol Ther 18(Suppl 1):14–23, PMID: 12925137
12. Houghton J, Stoicov C, Nomura S, Rogers AB et al (2004) Gastric cancer originating from bone marrow-derived cells. Science 306: 1568–1571, PMID: 15567866
13. Enno A, O'Rourke JL, Howlett CR et al (1995) MALToma-like lesions in the murine gastric mucosa after long-term infection with *Helicobacter felis* A mouse model of Helicobacter pylori-induced gastric lymphoma. Am J Pathol 147:217–222, PMID: 7604881
14. Enno A, O'Rourke J, Braye S, Howlett R, Lee A (1998) Antigen-dependent progression of mucosa-associated lymphoid tissue (MALT)-type lymphoma in the stomach Effects of antimicrobial therapy on gastric MALT lymphoma in mice. Am J Pathol 52:1625–1632, PMID: 9626066
15. Wang X, Willén R, Svensson M, Ljungh A, Wadström T (2003) Two-year follow-up of *Helicobacter pylori* infection in C57BL/6 and Balb/cA mice. APMIS 111:514–522, PMID: 12780527
16. Takata T, El-Omar E, Camorlinga M et al (2002) *Helicobacter pylori* does not require Lewis X or Lewis Y expression to colonize C3H/HeJ mice. Infect Immun 70:3073–3079, PMID: 12011000
17. Mähler M, Janke C, Wagner S, Hedrich HJ (2002) Differential susceptibility of inbred mouse strains to *Helicobacter pylori* infection. Scand J Gastroenterol 37:267–278, PMID: 11916188
18. Wang TC, Dangler CA, Chen D et al (2000) Synergistic interaction between hypergastrinemia and *Helicobacter* infection in a mouse model of gastric cancer. Gastroenterology 118:36–47, PMID: 10611152
19. Fox JG, Wang TC, Rogers AB et al (2003) Host and microbial constituents influence *Helicobacter pylori*-induced cancer in a murine model of hypergastrinemia. Gastroenterology 124:1879–1890, PMID: 12806621
20. Fox JG, Rogers AB, Ihrig M et al (2003) *Helicobacter pylori*-associated gastric cancer in INS-GAS mice is gender specific. Cancer Res 63:942–950, PMID: 12615707
21. Fox JG, Beck P, Dangler CA, Whary MT, Wang TC et al (2000) Concurrent enteric helminth infection modulates inflammation and gastric immune responses and reduces *helicobacter*-induced gastric atrophy. Nat Med 6:536–542, PMID: 10802709
22. Whary MT, Fox JG (2004) Th1-mediated pathology in mouse models of human disease is ameliorated by concurrent Th2 responses

to parasite antigens. Curr Top Med Chem 4:531–538, PMID: 14965304

23. Stoicov C, Whary M, Rogers AB et al (2004) Coinfection modulates inflammatory responses and clinical outcome of *Helicobacter felis* and *Toxoplasma gondii* infections. J Immunol 173:3329–3336, PMID: 15322196
24. Lofgren JL, Whary MT, Ge Z et al (2011) Lack of commensal flora in *H. pylori*-infected INS-GAS mice reduces gastritis and delays intraepithelial neoplasia. Gastroenterology 140:210–220, PMID: 20950613
25. Hasegawa S, Nishikawa S, Miura T et al (2004) Tumor necrosis factor-alpha is required for gastritis induced by *Helicobacter felis* infection in mice. Microb Pathog 37:119–124, PMID: 15351034
26. Sawai N, Kita M, Kodama T et al (1999) Role of gamma interferon in *Helicobacter pylori*-induced gastric inflammatory responses in a mouse model. Infect Immun 67:279–285, PMID: 9864227
27. Yamamoto T, Kita M, Ohno T et al (2004) Role of tumor necrosis factor-alpha and interferon-gamma in *Helicobacter pylori* infection. Microbiol Immunol 48:647–654, PMID: 15383700
28. El-Omar EM, Carrington M, Chow WH et al (2000) Interleukin-1 polymorphisms associated with increased risk of gastric cancer. Nature 404:398–402, PMID: 10746728
29. Tu S, Bhagat G, Cui G, Takaishi S et al (2008) Overexpression of interleukin-1beta induces gastric inflammation and cancer and mobilizes myeloid-derived suppressor cells in mice. Cancer Cell 14:408–419, PMID: 18977329
30. Berg DJ, Lynch NA, Lynch RG, Lauricella DM (1998) Rapid development of severe hyperplastic gastritis with gastric epithelial dedifferentiation in *Helicobacter felis*-infected IL-10 (−/−) mice. Am J Pathol 152:1377–1386, PMID: 9588906
31. Houghton JM, Bloch LM, Goldstein M, Von Hagen S, Korah RM (2000) In vivo disruption of the fas pathway abrogates gastric growth alterations secondary to *Helicobacter* infection. J Infect Dis 182:856–864, PMID: 10950781
32. Cai X, Stoicov C, Li H, Carlson J, Whary M, Fox JG, Houghton J (2005) Overcoming Fas-mediated apoptosis accelerates *Helicobacter*-induced gastric cancer in mice. Cancer Res 65:10912–10920, PMID: 16322238
33. Chu IM, Hengst L, Slingerland JM (2008) The Cdk inhibitor p27 in human cancer: prognostic potential and relevance to anticancer therapy. Nat Rev Cancer 8:253–267, PMID: 18354415
34. Shirin H, Sordillo EM, Kolevska TK et al (2000) Chronic *Helicobacter pylori* infection induces an apoptosis-resistant phenotype associated with decreased expression of p27 (kip1). Infect Immun 68:5321–5328, PMID: 10948161
35. Fero ML, Randel E, Gurley KE, Roberts JM, Kemp CJ (1998) The murine gene p27 Kip1 is haplo-insufficient for tumor supression. Nature 396:177–180, PMID: 9823898
36. Kuzushita N, Rogers AB, Monti NA et al (2005) p27 kip1 deficiency confers susceptibility to gastric carcinogenesis in *Helicobacter pylori*-infected mice. Gastroenterology 129: 1544–1556, PMID: 16285954
37. Ohnishi N, Yuasa H, Tanaka S et al (2008) Transgenic expression of *Helicobacter pylori* Cag A induces gastrointestinal and hematopoietic neoplasms in mouse. Proc Natl Acad Sci USA 105:1003–1008, PMID: 18192401
38. Hirayama F, Takagi S, Iwao E et al (1999) Development of poorly differentiated adenocarcinoma and carcinoid due to long-term *Helicobacter pylori* colonization in Mongolian gerbils. J Gastroenterol 34:450–454, PMID: 10452676
39. Ikeno T, Ota H, Sugiyama A, Ishida K, Katsuyama T, Genta RM, Kawasaki S (1999) *Helicobacter pylori*-induced chronic active gastritis, intestinal metaplasia, and gastric ulcer in Mongolian gerbils. Am J Pathol 154(3): 951–960, PMID: 10079274
40. Zheng Q, Chen XY, Shi Y et al (2004) Development of gastric adenocarcinoma in Mongolian gerbils after long-term infection with *Helicobacter pylori*. J Gastroenterol Hepatol 19:1192–1198, PMID: 15377299
41. Yoshizawa N, Takenaka Y, Yamaguchi H et al (2007) Emergence of spasmolytic polypeptide-expressing metaplasia in Mongolian gerbils infected with *Helicobacter pylori*. Lab Invest 87:1265–1276, PMID: 18004396
42. Watanabe T, Tada M, Nagai H et al (1998) *Helicobacter pylori* infection induces gastric cancer in mongolian gerbils. Gastroenterology 115:642–648, PMID: 9721161
43. Honda S, Fujioka T, Tokieda M et al (1998) Development of *Helicobacter pylori*-induced gastric carcinoma in Mongolian gerbils. Cancer Res 58:4255–4259, PMID: 9766647
44. Tatematsu M, Tsukamoto T, Mizoshita T (2005) Role of *Helicobacter pylori* in gastric carcinogenesis: the origin of gastric cancers and heterotopic proliferative glands in Mongolian gerbils. Helicobacter 10:97–106, PMID: 15810939
45. Peek RM Jr, Wirth HP, Moss SF et al (2000) *Helicobacter pylori* alters gastric epithelial cell

cycle events and gastrin secretion in Mongolian gerbils. Gastroenterology 118:48–59, PMID: 10611153
46. Israel DA, Salama N, Arnold CN, Moss SF et al (2001) *Helicobacter pylori* strain-specific differences in genetic ontent, identified by microarray, influence host inflammatory responses. J Clin Invest 107:611–620, PMID: 11238562
47. Correa P (1992) Human gastric carcinogenesis: a multistep and multifactorial process-First American Cancer Society Award Lecture on Cancer Epidemiology and Prevention. Cancer Res 52:6735–6740, PMID: 1458460
48. Wroblewski LE, Peek RM Jr, Wilson KT (2010) *Helicobacter pylori* and gastric cancer: factors that modulate disease risk. Clin Microbiol Rev 23:713–739, PMID: 20930071
49. Sugiyama A, Maruta F, Ikeno T et al (1998) *Helicobacter pylori* infection enhances N-methyl-N-nitrosourea-induced stomach carcinogenesis in the Mongolian gerbil. Cancer Res 58:2067–2069, PMID: 9605743
50. Tokieda M, Honda S, Fujioka T et al (1999) Effect of *Helicobacter pylori* infection on the N-methyl-N′-nitro-N-nitrosoguanidine-induced gastric carcinogenesis in mongolian gerbils. Carcinogenesis 20:1261–1266, PMID: 10383899
51. Yamachika T, Nakanishi H, Inada K et al (1998) *N*-methyl-*N*-nitrosoureaconcentration-dependent, rather than total intake-dependent, induction of adenocarcinomas in the glandular stomach of BALB/c mice. Jpn J Cancer Res 89:385–391, PMID: 9617343
52. Han SU, Kim YB, Joo HJ et al (2002) *Helicobacter pylori* infection promotes gastric carcinogenesis in a mice model. J Gastroenterol Hepatol 17:253–261, PMID: 11982694
53. Fuccio L, Zagari RM, Eusebi LH et al (2009) Meta-analysis: can *Helicobacter pylori* eradication treatment reduce the risk for gastric cancer? Ann Intern Med 151:121–128, PMID: 19620164
54. Lee CW, Rickman B, Rogers AB et al (2008) *Helicobacter pylori* eradication prevents progression of gastric cancer in hypergastrinemic INS-GAS mice. Cancer Res 68:3540–3548, PMID: 18441088
55. Romero-Gallo J, Harris EJ, Krishna U et al (2008) Effect of *Helicobacter pylori* eradication on gastric carcinogenesis. Lab Invest 88: 328–336, PMID: 18180700
56. Zullo A, Hassan C, Cristofari F et al (2010) Effects of *Helicobacter pylori* eradication on early stage gastric mucosa-associated lymphoid tissue lymphoma. Clin Gastroenterol Hepatol 8:105–110, PMID: 19631287
57. Lee CW, Rickman B, Rogers AB et al (2009) Combination of sulindac and antimicrobial eradication of *Helicobacter pylori* prevents progression of gastric cancer in hypergastrinemic INS-GAS mice. Cancer Res 69:8166–8174, PMID: 19826057
58. DeLyria ES, Redline RW, Blanchard TG (2009) Vaccination of mice against *H. pylori* induces a strong Th-17 response and immunity that is neutrophil dependent. Gastroenterology 247–256, PMID: 18948106

Chapter 13

Bacterial Culture and Inoculation of Mice (Simple Infection)

Brian M. Gray and Kathryn A. Eaton

Abstract

Standardization of bacterial culture is crucial for in vivo experiments addressing Helicobacter/host interaction. Here we present methods for bacteria culture and infection of mice.

Key words: Simple infection, Animal models, In vitro bacterial culture

1. Introduction

Mice were first successfully colonized with a disease-producing strain of *H. pylori* in 1997. Since that time, a number of different mouse models have been described utilizing different mouse and bacterial strains and inoculation protocols, as well as different methods of outcome assessment. In this chapter, we describe methods of bacterial inoculation. In subsequent chapters we describe adoptive transfer of lymphocytes and tissue collection and processing, including isolation of bacteria, DNA, RNA, and splenic and gastric immune cells for analysis. In addition, we address differences in mouse and bacterial strains, and discuss morphologic evaluation of mouse models in the context of human disease.

The laboratory mouse is currently by far the animal model most commonly used for the study of human disease. Mice are small with a rapid generation time and are genetically manipulable, resulting in dozens of available inbred strains and genetically engineered mutants designed to evaluate specific mammalian genes in the context of infectious or other diseases. Immunological reagents that react with mouse tissues are readily available, and reproducible methods are available for modeling many diseases in mice. In spite of the overall utility of mice as models of human disease, there are a number of factors that must be considered when interpreting experimental

JeanMarie Houghton (ed.), *Helicobacter Species: Methods and Protocols*, Methods in Molecular Biology, vol. 921, DOI 10.1007/978-1-62703-005-2_13,

results of mouse models. Because they are mammals, mice are similar to humans physiologically and in many features of their immune responses. However, specific differences between mice and humans must be taken into account when interpreting experimental results. Also, mice, like other nonhuman mammals, are not always susceptible to infection by human pathogens, and when infection occurs, the manifestations of disease are rarely, if ever, exactly the same as they are in human patients. Infection of mice with cognate mouse pathogens can sometimes circumvent these issues, but may introduce additional interpretive difficulties because of physiologic and molecular differences between the mouse and human pathogens themselves. For example, mice are more susceptible to gastritis due to a related organism, *Helicobacter felis*, than they are to *H. pylori*, making *H. felis* a useful tool in investigation of host immune responses to gastric helicobacters (1) as well as preneoplastic changes (2). However, *H. felis* lacks many bacterial genes and functions that are of interest in the pathogenesis of disease due to *H. pylori* (most prominently the *cag* pathogenicity island, a series of genes that is epidemiologically associated with severity of disease) (3), limiting the usefulness of *H. felis* for evaluation of specific bacterial virulence factors. Thus, use and interpretation of mouse models of human disease must be carried out and interpreted with a clear knowledge and understanding of the differences between mice and humans and/or the differences between mouse and human pathogens. That said, mouse models of human disease have been and are increasingly central in our ability to understand basic mechanisms of disease in mammals, and can form the basis for investigation and study of disease in the natural human host.

Prior to 1997, the principal animal models used for study of disease due to *H. pylori* were germ-free piglets (4, 5) and nonhuman primates (6). While these models were pivotal in identifying *H. pylori* virulence determinants and the basis of host immune responses, they were expensive, and their use was limited to those investigators who had access to the necessary institutional resources. A few published studies reporting colonization of mice by *H. pylori* had been published, but colonization rates were low, disease manifestations were nonexistent, and results were generally discouraging (7–10). In 1997 Lee et al. described in detail a mouse model of *H. pylori* in which a mouse-adapted strain, originally called HP8 and later called SS1, was used (11). The strain was isolated from a human patient and then passed in mice resulting in consistently high rates of colonization (10^7 cfu/g of stomach), colonization of several inbred mouse strains, and chronic progressive gastritis with many features that were similar to the most common manifestations of disease in humans: slowly progressive and subclinical chronic or chronic active gastritis. Thus, strain SS1 became the most commonly used strain for mouse models, and is the strain that is most commonly used by our laboratory (for comments on mouse-adapted *H. pylori* strains, see Note 1).

2. Types and Uses of Mouse Models

Current use of mouse models of *H. pylori* may be roughly divided into two categories, referred to here as "simple infection" and "adoptive transfer" (see Chapter 14). Simple infection involves oral inoculation of mice with live, broth-cultured *H. pylori*. C57BL/6J mice or congenic mutant strains are most commonly used (see Note 2), and infection results in mild, slowly progressive chronic or chronic active gastritis that is generally detectable by 8–12 weeks after inoculation, and may become severe in some mice by 6 months–1 year (12–18). This model is most useful for examination of bacterial colonization factors since gastric lesions are not present in until 8–12 weeks after inoculation, and before that time, histologic examination is unrewarding.

3. Bacterial Culture and Inoculation of Mice (Simple Infection)

3.1. Materials

Buffers and media: All solutions are made with deionized distilled water (DDW) and handled aseptically. They are autoclaved or sterilized by filtration through a 0.22 μm filter, unless otherwise indicated. Unless otherwise noted, all supplies and reagents are from Fisher Scientific or Sigma Aldrich.

3.2. Brucella broth (Difco) with 10% fetal calf serum (BB+FCS)

- 7 g Brucella broth powder (Difco).
- 225 ml DDW.
- Autoclave 20 min, cool.
- Add 25 ml heat-inactivated FCS.

3.3. Urease indicator broth

- 5 ml 0.2 M $NaPO_4$ buffer, pH 6.5 (23 g $NaH_2PO_4 \cdot 2H_2O$ + 14.2 g $Na_2HPO_4 \cdot 12H_2O$/L, adjust pH with HCl or NaOH).
- 6.6 ml 5 M urea (30 g/100 ml).
- 20 ml 0.1% Phenol Red.
- 1.0 ml 2% NaN_3 (20 mg/100 ml).
- QS to 100 ml with DDW.

3.4. Modified Skirrow's antibiotic supplement (see Note 3)

- Aseptically prepare the following solutions:
- Vancomycin: 200 mg in 1 ml of DDW.
- Polymyxin: 6.6 mg in 1 ml of DDW.
- Trimethoprim: 50 mg in 1 ml of DDW.
- Bacitracin: 400 mg in 15 ml of DDW.
- Amphotericin B: 100 mg in 5 ml of 1 M NaOH.
- Nalidixic Acid: 20.2 mg in 1 ml of 1 M NaOH.

Filter sterilize separately into a single 50 ml conical tube, adding the naladixic acid last. Suspension will become slightly cloudy. QS to 25 ml with DDW. Cover tube with foil and store at 4°C protected from light. Vortex before using and dilute 250 μl of solution/20 ml broth to achieve the following final concentrations (see Note 4):

Antibiotic	Concentration of stock solution	Final concentration
Vancomycin	8 mg/ml	100 μg/ml
Polymyxin	0.264 mg/ml	3.3 μg/ml
Trimethoprim	2 mg/ml	25 μg/ml
Bacitracin	16 mg/ml	200 μg/ml
Amphotericin B	0.8 mg/ml	50 μg/ml
Nalidixic acid	4 mg/ml	10 μg/ml

3.5. Selective plates for marked strain

- We use kanamycin and chloramphenicol for marking mutant *H. pylori* strains.
- Both are used at 20 μg/ml final concentration.
- Kanamycin (10 mg/ml) is purchased from Sigma.
- Chloramphenicol stock solution is made in 100% ethanol at 10 mg/ml. Both are stored at 4°C and diluted 40 μl per plate or per 20 ml broth (see Note 4).

3.6. Other supplies

- TSA sheep blood agar plates treated with modified Skirrow's antibiotic supplement (see Note 4).
- Tissue culture plates.
- 15 ml and 50 ml conical tubes.
- Hemocytometer.

4. Methods

4.1. Bacterial Culture and Primary Inoculation of Mice (Simple Infection)

Mice are inoculated with broth-cultured *H. pylori* SS1 in mid-logarithmic phase growth:

1. Place 10 ml BB + FCS in a sterile 100 mm sterile plastic Petri dish.
2. Inoculate with 0.1 ml of thawed frozen bacterial stock or 1 loopful from a 2-day-old lawn of *H. pylori* SS1 (see Note 5).
3. Incubate overnight in a triple gas incubator (10% CO_2, 5% O_2, 85% N) at 37°C with gentle agitation (see Note 6).

4. After 18–24 h of incubation, examine the bacteria by light microscopy to evaluate motility and morphology, and perform urease and catalase tests. Bacteria for inoculation should be about 10^8cfu/ml, and actively motile with no clumping.
5. For urease and catalase tests, place two drops of culture on a glass slide and add a drop of urease broth to one and H_2O_2 to the other. Urease broth should turn from orange to bright cherry red in less than 5 min. H_2O_2 should immediately bubble.
6. If bacteria are present and motile, but not yet in late logarithmic growth (i.e., less than 10^8 cfu/ml), recheck the broth after 6–12 additional hours of incubation.
7. If the culture is sufficiently grown, quantify an aliquot using a standard laboratory hemocytometer.
8. Dilute the broth to approximately 10^7 cfu/ml (use NBF for dilution if bacteria are extremely motile).
9. Count manually by direct examination under 400× magnification. We find that this method is more accurate than optical density since optical density readings are affected by the age and morphology of the cultures. For viable plate count, seven tenfold dilutions on TSA sheep blood agar plates as described below.
10. Prepare the inoculum by centrifugation in a tabletop centrifuge at 2,500 × *g* for 20 min, and resuspend the bacterial colonies in sterile BB + FCS. Our standard protocol is to resuspend the bacteria at approximately 10^7 cfu/ml and orally inoculate mice with 0.1 ml (10^6 cfu). We have found that concentration is not critical, however, since mice are susceptible to colonization by as few as 200 cfu of *H. pylori* strain SS1 (unpublished data).
11. For inoculation, restrain the mice manually. Mice are held vertically (nose up) to prevent bending of the neck during inoculation. The inoculum is placed in a 1 cc syringe with an 18 g ball-tipped bent gavage needle, which is gently introduced into the esophagus. Care must be taken to place the tube without traumatizing the tissue or compromising the respiratory tract. Inexperienced individuals should obtain training prior to attempting this procedure. Once the gavage tube is within the esophagus (beyond the pharynx), 0.1 ml of bacterial suspension is slowly introduced and allowed to flow into the stomach. The tube is removed and the mouse held vertically briefly to allow complete swallowing. Mice should be observed for signs of difficulty during and after the procedure.
12. It is not necessary to fast the mice or administer acid-reducing treatment prior to bacterial inoculation. If desired, the inoculation may be repeated once or twice daily or every other day, but this is not necessary.

5. Notes

1. *Mouse-adapted H. pylori*. Recent studies have revealed that the early failure of mouse models noted above was most likely attributable to the natural phenotypic variability of human *H. pylori* isolates. Several studies have now indicated that regardless of mouse passage, bacterial virulence factors, or other considerations examined, about 20–40% of *H. pylori* isolates are able to colonize mice (19–23). Several strains have been passaged in mice, sometimes resulting in higher colonization density by the passaged isolate compared to the original clinical isolate (20, 24, 25). These strains are sometimes used in preference to SS1 because of ease of genetic manipulation, presence of specific *H. pylori* genes, or other factors (15, 26, 27).

   It should be noted that the use of SS1 has generated controversy because of the failure of that strain to elicit IL-8 production by cultured AGS cells (22, 27, 28). This has led to the suggestion that the cag-PAI is dysfunctional in SS1, and/or that the dysfunction is associated with mouse passage and is permissive for mouse colonization. However, as noted above, other studies have failed to demonstrate an association between the cag-PAI and mouse passage or colonization (19, 20, 23, 25). As a general rule mouse-colonizing strains optimally should colonize C57BL/6J mice at about 10^7 cfu/g of stomach wall and be visible in Warthin–Starry or Steiner-stained histologic sections in a multifocal distribution in the gastric mucosa. Colonization may be higher in the absence of gastritis (for example, in immunodeficient mice) (12) and may differ in different mouse strains (11, 29–31).
2. *Mouse strains*. C57BL/6 mice and mutant strains on this background are the strain most commonly used for studies of *H. pylori*. Most studies that compared strains showed that C57BL/6 mice supported strong colonization and developed more severe gastritis than the other strains tested (11, 29–31). It should be noted, however, that both genetics and normal enteric microbiota may vary depending on the source and subline of the mice, and could result in different responses to infection by identical mouse strains (32). Our laboratory uses helicobacter-free C57BL/6J mice and mutant strains purchased from Jackson laboratories. We find that these mice respond well and produce consistent results.
3. *Use of antibiotics*. In our laboratory we use antibiotics in bacterial media only when recovering bacteria from mouse stomachs or to isolate antibiotic-marked bacterial mutants. We find that antibiotics are not necessary to prevent contamination of inocula. Therefore, we prefer to avoid antibiotic use, thereby minimizing to unnecessary selection pressure on

laboratory-passaged strains. For recovery of bacteria from contaminated samples (i.e., mouse stomachs), we find the modified Skirrow's mixture sufficient for most samples. Some strains of mouse stomach bacteria are able to grow in the presence of these antibiotics. However, complete removal of the squamous mucosa prior to homogenization and plating of the tissue usually removes enough of these organisms to permit quantification of *H. pylori* colonies (see necropsy technique, below). The appearance of *H. pylori* colonies is sufficiently characteristic to distinguish them from contaminants. Should there be any doubt, individual colonies may be rapidly identified as *H. pylori* by collecting a sample on a sterile cotton swab and placing a drop of urease indicator broth on the swam. If the sample is *H. pylori*, the swab will turn cherry red within a few minutes.

4. *Antibiotic-treated agar plates.* As an alternative to including antibiotics in the plates during pouring, antibiotics may be added to prepurchased plates. Pipet the appropriate volume on the plate, spread the solution over the entire surface, and allow the plate to dry before use.
5. *Inoculum preparation.* For mouse inoculation it is important to pass bacteria as lawns rather than select single colonies. Bacteria, particularly *H. pylori*, undergo rapid and often undetectable genetic variation when grown in culture. Isolation of single colonies fixes these variations and may be associated with marked changes in colonization ability and/or disease induction. We have found that isolates of either SS1 or M6, another mouse-colonizing strain, range in their colonization fitness from 0 to 10^7 cfu/g of gastric mucosa after cloning and isolation of single colonies (unpublished observations). For that reason, we use bacterial lawns for inoculation, and minimize in vitro passages of our inoculum strains.
6. *Microaerobic atmosphere.* If a triple gas incubator is not available, a tissue-culture incubator set at 10% CO_2 in air or a CampyPak (BD diagnostic) will suffice.

References

1. Lee A, Hazell SL, O'Rourke J, Kouprach S (1988) Isolation of a spiral-shaped bacterium from the cat stomach. Infect Immun 56:2843–2850
2. Rogers AB, Fox JG (2004) Inflammation and Cancer. I. Rodent models of infectious gastrointestinal and liver cancer. Am J Physiol Gastrointest Liver Physiol 286:G361–366
3. Censini S, Lange C, Xiang Z, Crabtree JE, Ghiara P, Borodovsky M, Rappuoli R, Covacci A (1996) *Cag*, a pathogenicity island of *Helicobacter pylori*, encodes type I-specific and disease-associated virulence factors. Proc Natl Acad Sci U S A 93:14648–14653
4. Krakowka S, Morgan DR, Kraft WG, Leunk RD (1987) Establishment of gastric *Campylobacter pylori* infection in the neonatal gnotobiotic piglet. Infect Immun 55:2789–2796
5. Lambert JR, Borromeo M, Pinkard KJ, Turner H, Chapman CB, Smith ML (1987)

Colonization of gnotobiotic piglets with *Campylobacter Pyloridis*—an animal model? J Infect Dis. 155:1344

6. Bronsdon MA, Schoenknecht FD (1988) *Campylobacter pylori* isolated from the stomach of the monkey *Macaca nemestrina*. J Clin Microbiol 26:1725–1728
7. Ehlers S, Warrelmann M, Hahn H (1988) In search of an animal model for experimental *Campylobacter pylori* infection: administration of *Campylobacter pylori* to rodents. Zentralblatt Fur Bakteriologie 268:341–346
8. Cantorna MT, Balish E (1990) Inability of human clinical strains of *Helicobacter pylori* to colonize the alimentary tract of germfree rodents. Can J Microbiol 36:237–241
9. Karita M, Kouchiyama T, Okita K, Nakazawa T (1991) New small animal model for human gastric *Helicobacter pylori* infection: success in both nude and euthymic mice. Am J Gastroentero 86:1596–1603
10. Marchetti M, Arico B, Burroni D, Figura N, Rappuoli R, Ghiara P (1995) Development of a mouse model of *Helicobacter pylori* infection that mimics human disease. Science 267: 1655–1658
11. Lee A, Orourke J, Deungria MC, Robertson B, Daskalopoulos G, Dixon MF (1997) A standardized mouse model of *Helicobacter pylori* infection: introducing the Sydney strain. Gastroenterology 112:1386–1397
12. Eaton KA, Ringler SR, Danon SJ (1999) Murine splenocytes induce severe gastritis and delayed-type hypersensitivity and suppress bacterial colonization in *Helicobacter pylori*-infected SCID mice. Infect Immun 67:4594–4602
13. Eaton KA, Mefford M, Thevenot T (2001) The role of T cell subsets and cytokines in the pathogenesis of *Helicobacter pylori* gastritis in mice. J Immunol 166:7456–7461
14. Eaton KA, Mefford ME (2001) Cure of *Helicobacter pylori* infection and resolution of gastritis by adoptive transfer of splenocytes in mice. Infect Immun 69:1025–1031
15. Eaton KA, Gilbert JV, Joyce EA, Wanken AE, Thevenot T, Baker P, Plaut A, Wright A (2002) In vivo complementation of *ureB* restores the ability of *Helicobacter pylori* to colonize. Infect Immun 70:771–778
16. Eaton KA, Logan SM, Baker PE, Peterson RA, Monteiro MA, Altman E (2004) *Helicobacter pylori* with a truncated lipopolysaccharide O chain fails to induce gastritis in SCID mice injected with splenocytes from wild-type C57BL/6J mice. Infect Immun 72: 3925–3931
17. Eaton KA, Benson LH, Haeger J, Gray BM (2006) Role of transcription factor T-bet expression by CD4+ cells in gastritis due to *Helicobacter pylori* in mice. Infect Immun 74:4673–4684
18. Eaton KA, Danon SJ, Krakowka S, Weisbrode SE (2007) A reproducible scoring system for quantification of histologic lesions of inflammatory disease in mouse gastric epithelium. Comp Med 57:57–65
19. Ayraud S, Janvier B, Fauchere JL (2002) Experimental colonization of mice by fresh clinical isolates of *Helicobacter pylori* is not influenced by the *cagA* status and the *vacA* genotype. FEMS Immunol Med Microbiol 34:169–172
20. Thompson LJ, Danon SJ, Wilson JE, O'Rourke JL, Salama NR, Falkow S, Mitchell H, Lee A (2004) Chronic *Helicobacter pylori* infection with Sydney strain 1 and a newly identified mouse-adapted strain (Sydney strain 2000) in C57BL/6 and BALB/c mice. Infect Immun 72:4668–4679
21. Chionh YT, Walduck AK, Mitchell HM, Sutton P (2009) A comparison of glycan expression and adhesion of mouse-adapted strains and clinical isolates of *Helicobacter pylori*. FEMS Immunol Med Microbiol 57:25–31
22. Philpott DJ, Belaid D, Troubadour P, Thiberge JM, Tankovic J, Labigne A, Ferrero RL (2002) Reduced activation of inflammatory responses in host cells by mouse-adapted *Helicobacter pylori* isolates. Cell Microbiol 4:285–296
23. Suresh MR, Fanta MB, Kriangkum J, Jiang Q, Taylor DE (2000) Colonization and immune responses in mice to *Helicobacter pylori* expressing different Lewis antigens. J Pharm Pharm Sci 3:259–266
24. Shi Y, Liu XF, Zhuang Y, Zhang JY, Liu T, Yin Z, Wu C, Mao XH, Jia KR, Wang FJ, Guo H, Flavell RA, Zhao Z, Liu KY, Xiao B, Guo Y, Zhang WJ, Zhou WY, Guo G, Zou QM (2010) *Helicobacter pylori*-induced Th17 responses modulate Th1 cell responses, benefit bacterial growth, and contribute to pathology in mice. J Immunol 184:5121–5129
25. Salaun L, Ayraud S, Saunders NJ (2005) Phase variation mediated niche adaptation during prolonged experimental murine infection with *Helicobacter pylori*. Microbiology 151: 917–923
26. Van Doorn NEM, Namavar F, Sparrius M, Stoof J, Vanrees EP, Vandoorn LJ, Vandenbrouckegrauls CMJE (1999) *Helicobacter pylori*—associated gastritis in mice is host and strain specific. Infect Immun 67:3040–3046

27. Crabtree JE, Ferrero RL, Kusters JG (2002) The mouse colonizing *Helicobacter pylori* strain SS1 may lack a functional cag pathogenicity island. Helicobacter 7:139–140, discussion 140–131
28. van Doorn NE, Namavar F, Sparrius M, Stoof J, van Rees EP, van Doorn LJ, Vandenbroucke-Grauls CM (1999) *Helicobacter pylori*-associated gastritis in mice is host and strain specific. Infect Immun 67:3040–3046
29. Dey A, Yokota K, Kosavashi K, Oguma K, Hirai Y, Akagi T (1998) Antibody and cytokine responses in *Helicobacter pylori*-infected various mouse strains. Acta Medica Okayama 52:41–48
30. Kamradt AE, Greiner M, Ghiara P, Kaufmann SH (2000) *Helicobacter pylori* infection in wild-type and cytokine-deficient C57BL/6 and BALB/c mouse mutants. Microbes Infect 2:593–597
31. Smythies LE, Chen J, Lindsey JR, Ghiara P, Smith PD, Waites KB (2000) Quantitative analysis of *Helicobacter pylori* infection in a mouse model. J Immunol Methods 242:67–78
32. Ivanov II, Littman DR (2010) Segmented filamentous bacteria take the stage. Mucosal Immunol 3:209–212

Chapter 14

Adoptive Transfer of Splenocytes to Immunocompromised Mice

Brian M. Gray and Kathryn A. Eaton

Abstract

Analysis of the immune response of mice to Helicobacter infection has been greatly aided by the use of various deficient mouse strains. Here we present protocols for reconstitution of immune-deficient mice with wild-type immune cells and protocols for analysis of the outcome.

Key words: Splenocytes, Immunodeficient, Wild type

1. Introduction

The adoptive transfer model involves oral inoculation of immunodeficient mice (either severe, combined, immunodeficient [SCID] or recombination-activating gene knockout [RAG KO] mice) with *Helicobacter pylori*, and subsequent adoptive transfer of immune cells from congenic immunocompetent mice. The most common protocol (and the one used in our laboratory) is adoptive transfer of splenic CD4+ T cells 2 weeks after bacterial infection. Recipients of CD4+ splenocytes from C57BL/6J mice develop rapidly progressive severe gastritis that is detectable within 2 weeks of transfer, and within 8–10 weeks progresses to severe chronic active inflammation, epithelial erosions, epithelial hyperplasia, and mucus metaplasia (1). Because disease is more severe and more rapidly progressive in this model than in the simple infection model, it is useful for examination of host immune response and disease, and the interaction of specific bacterial and host immune factors in disease manifestations.

JeanMarie Houghton (ed.), *Helicobacter Species: Methods and Protocols*, Methods in Molecular Biology, vol. 921, DOI 10.1007/978-1-62703-005-2_14,

2. Materials

2.1. General

See Chapter 13 for growth of Hp and infection of mice.

1. Forceps, surgical scissors, and scalpel with sterile blades for necropsy and tissue collection.
2. 6-Well tissue culture plates.
3. 15 and 50 ml conical tubes.
4. 40 μm Mesh nylon cell strainers.
5. Magnetic bead separation supplies (Miltenyi Biotec, Auburn, CA):
   - Antibody-coated magnetic microbeads (specifically L3T4 anti-mouse CD4 beads).
   - Magnet and stand.
   - MS separation columns.
6. 4 ml Test tubes for fluorescent antibody staining and flow cytometry.

2.2. 10% RPMI for splenocyte isolation

- 500 ml RPMI-1640.
- 50 ml Heat-inactivated fetal calf serum.
- 5 ml Penicillin/streptomycin (10,000 units penicillin and 10 mg streptomycin/ml, Sigma)

2.3. MACS bead buffer

- 1× PBS.
- 0.5% w/v BSA.
- 2 mM EDTA.
- Filter sterilize and store at 4°C.

3. Methods

Splenocytes are isolated from spleens of 4–6-week-old C57BL/6J mice for transfer into congenic SCID or RAG KO mice. Donor and recipient mice are of the same sex. At least two donor mice are used per transfer. For CD4 T cell transfer, two donors are sufficient to inoculate up to 10 recipient mice.

3.1. Methods for Splenocyte isolation and fractionation (see Note 1)

1. Euthanize a healthy 4–6-week-old C57BL/6J mouse as described in the chapter on euthanasia and necropsy (Chapter 17).
2. Aseptically remove the spleen as described in the chapter on euthanasia and necropsy (Chapter 17) and place in 3 ml of ice-cold 10% RPMI in a sterile Petri dish.

3. Using the ridged end of a plunger from a 3 cc syringe, mash the spleen until there are no remaining chunks of tissue.
4. Filter the single-cell suspension through a 40 μm nylon mesh cell strainer into a 50 ml conical tube. Rinse the Petri dish with another 10–12 ml RPMI and filter the rinse into the same tube.
5. Pellet the splenocytes by centrifugation at 450 × *g* for 5 min at 4°C.
6. Discard the supernatant and resuspend the cells in 450 μl of cold MACS bead buffer.
7. Mix the cells with 50 μl L3T4 anti-mouse CD4 magnetic microbeads. Incubate for 15 min at 4°C.
8. During the cell–bead incubation, insert the separation column into the magnet, and pre-wet it with 0.5 ml MACS bead buffer.
9. Transfer the cell–bead mixture to the separation column and allow it to flow through.
10. Wash the column twice with 0.5 ml of cold MACS bead buffer.
11. Remove the column from the magnet. Flush the CD4+ cells from the column into a 15 ml conical tube by adding 1 ml of MACS bead buffer and applying the plunger supplied with the column.
12. Wash cells twice in 1× PBS with centrifugation at 450 × *g* for 5 min at 4°C. Resuspend in a final volume of 1 ml 1× PBS.
13. Count the cells and bring to a final concentration of 1×10^7 cells/ml in 1× PBS.
14. For adoptive transfer, inject 0.1 ml per mouse into the peritoneal cavity. For this, restrain the mouse in dorsal recumbancy. Using a 1 cc syringe and 25 g needle, slowly introduce the needle through the skin and body wall into the caudal abdominal cavity, and inject the cell suspension (see Note 2).

4. Notes

1. *Splenocyte fractionation.* CD4+ cells are both necessary and sufficient to induce gastritis in *H. pylori*-infected recipient mice, although transfer of unfractionated splenocytes results in similar lesions and immune responses (1–3). We currently use CD4+ T cell fractions for adoptive transfer, and the methods included here are limited to CD4+ cells. This magnetic bead isolation method may be used to enrich for specific CD4+ cell subsets

through positive selection (for example, CD4+CD25+ regulatory cells) or to select other cell subsets. In addition, CD4+ cells may be isolated by negative selection using the CD4+ T cell Isolation Kit II beads (Miltenyi Biotec).

2. *Splenocyte transfer*. Intravenous injection of splenocytes may be used but is not necessary. We find that intraperitoneal injection as described results in engraftment of CD4+ cells in the recipient mice typically amounting to 7–12% of total splenocytes by 8 weeks after transfer.

References

1. Eaton KA, Ringler SR, Danon SJ (1999) Murine splenocytes induce severe gastritis and delayed-type hypersensitivity and suppress bacterial colonization in *Helicobacter pylori*-infected SCID mice. Infect Immun 67:4594–4602
2. Eaton KA, Mefford M, Thevenot T (2001) The role of T cell subsets and cytokines in the pathogenesis of *Helicobacter pylori* gastritis in mice. J Immunol 166:7456–7461
3. Eaton KA, Mefford ME (2001) Cure of *Helicobacter pylori* infection and resolution of gastritis by adoptive transfer of splenocytes in mice. Infect Immun 69: 1025–1031

Chapter 15

Isolation of Gastric Lamina Propria Leukocytes

Brian M. Gray and Kathryn A. Eaton

Abstract

Immune cells recruited to the infected gastric mucosa can be isolated and used for a variety of purposes. Here we describe methods for the isolation and characterization of gastric lamina propria leukocytes.

Key words: Local immune environment, Leukocytes, Gastric mucosa

1. Introduction

The systemic immune response to Helicobacter infection may not be reflective of the local immune response. It is essential to understand the cellular composition of the local immune environment in response to Helicobacter infection. In order to do this, leukocytes are isolated directly from the gastric lamina propria of infected mice (see Note 1).

2. Materials

Buffers for gastric lamina propria leukocyte (LPL) isolation: Adapted from Lefrancois and Lycke (1).

2.1. Solution #1

1. 50 ml 1× Hank's buffered salt solution (HBSS, Gibco) (Ca^{++} and Mg^{++} free).
2. 7.5 ml 1 M HEPES (Gibco).
3. 250 ml DDW.

JeanMarie Houghton (ed.), *Helicobacter Species: Methods and Protocols*, Methods in Molecular Biology, vol. 921, DOI 10.1007/978-1-62703-005-2_15, © Springer Science+Business Media, LLC 2012

4. Adjust pH to 7.2 with NaOH.
5. QS to 500 ml with DDW.
6. Store for no more than 2 weeks at 4°C.

2.2. Solution #2

1. 45 ml 10× HBSS (Ca^{++} and Mg^{++} free).
2. 50 ml Fetal calf serum.
3. 7.5 ml 1 M HEPES.
4. 10 ml 0.25 M EDTA.
5. 250 ml DDW.
6. Adjust pH to 7.2 with NaOH.
7. DDW to 500 ml.

2.3. RPMI-10 for LPL

1. 45 ml 10× RPMI-1640 without sodium bicarbonate (Gibco).
2. 50 ml Fetal calf serum.
3. 7.5 ml 1 M HEPES.
4. 250 ml DDW.
5. Adjust pH to 7.2 ml with NaOH.
6. DDW to 500 ml.
7. Prepare fresh for each isolation.

2.4. RPMI-10 with Liberase

Immediately prior to use, add 1.73 ml DDW to a 5-mg vial of Liberase-TM (Roche Applied Science) and dissolve. The final concentration of enzyme is 15 Wünsch U/ml. Add 1 ml of reconstituted Liberase-TM to 100 ml of RPMI-10, above (see Note 2).

3. Methods

1. As described below (Chapter 17) remove the stomach from an infected, adoptively transferred RAG KO or SCID mouse, place in a sterile Petri dish, open, and gently remove the contents.
2. Collect two longitudinal strips from the greater curvature for histologic examination, and then remove the squamous mucosa and duodenum as described below (Chapter 17).
3. If splenic lymphocytes are to be analyzed, remove the spleen from the mouse, and place it in one well of a 6-well plate with 3 ml of RPMI-10. Keep the plate on ice until ready to process the spleen(s).
4. Transfer the remaining gastric tissue mucosa-side down into one well of a 6-well plate with 3 ml of Solution #1. Keep the plate on ice until all tissues are collected.

5. Pre-label two 50 ml conical tubes with an identifying number for each tissue sample. Then mark one tube to be used for washes and digestions ("processing"), and label the other for "collection." All cells harvested from one sample will be collected into the latter tube.
6. Gently wash the mucosa of the stomachs by picking up the muscularis side of the tissue with forceps and agitating the tissue in the well.
7. Transfer the tissue to a Petri dish and place it muscularis-side down. With a scalpel, dice the tissue into pieces approximately 1 mm square.
8. Transfer the tissue fragments into the labeled processing tube and add 10 ml of Solution #1.
9. Gently agitate the tissue on an orbital shaker or multi-tube rotator for 5 min at room temperature.
10. Pour the suspension through a nylon cell strainer into the labeled collection tube. Keep collected, pooled cells on ice.
11. Rinse the tissue fragments from the strainer and back into the original processing tube, using 5–10 ml of Solution #2 as needed. Combine the washes and bring the final volume to 15 ml (see Note 3).
12. Shake the tissue vigorously (approximately 100 rpm on an orbital shaker) for 15 min at 37°C.
13. Pour the suspension through the nylon cell strainer into the collection tube.
14. Repeat steps 11 through 13 once.
15. Rinse the tissue fragments off the strainer and back into the processing tube, now using 5–10 ml of RPMI-10. Bring the final volume to 10 ml with RPMI-10.
16. Shake the tissue vigorously for 15 min at room temperature. (This wash is necessary to remove lingering EDTA from Solution #2, which may interfere with protease digestion.)
17. During this incubation, centrifuge the collected cells at $850 \times g$ for 5 min at 4°C. Gently pour off or aspirate the supernatant as the cells are only loosely pelleted. Leave approximately 5 ml of supernatant on the cells.
18. Pour the suspension in the processing tube (step 16) through the nylon cell strainer into the collection tube.
19. Rinse the tissue fragments off the strainer and back into the processing tube using 5–10 ml of RPMI-10 with Liberase. Bring the final volume to 15 ml with RPMI-10 + Liberase.
20. Digest the tissue with vigorous shaking for 45 min at 37°C.
21. Repeat steps 18 through 20 once.

22. During the second digestion, process the collected spleens as described above (see splenocyte Isolation) in RPMI-10. Transfer 0.5 ml into test tubes for fluorescent antibody staining and flow cytometry.
23. At the end of the second digestion, pour the suspension through the nylon cell strainer into the collection tube. Discard the strainer.
24. Centrifuge the collected cells at $850 \times g$ for 5 min at 4°C. Gently pour off or aspirate the supernatant.
25. Resuspend the cells in 5 ml of RPMI-10. Transfer 0.5 ml into test tubes for fluorescent antibody staining and flow cytometry.

4. Notes

1. *Gastric LPL yield.* Normal uninflamed mouse stomach contains few leukocytes in the mucosa, and attempts to isolate gastric LPL are unrewarding. However, when severe gastritis is present (i.e., in infected recipient mice 8 weeks after transfer) sufficient numbers of leukocytes are present to allow isolation and identification by flow cytometry. Most of the stomach is required for this procedure, precluding additional assays other than histologic confirmation of gastritis. Our average viable cell yield from one mouse stomach is 2.5×10^6 cells total. This population comprises approximately 20% leukocytes of which most are CD4+.
2. *Liberase digestion.* This protocol calls for 30 ml of RPMI-10 w/Liberase per mouse stomach. Dilute enough enzyme for the number of mice to be processed.
3. *LPL washes.* We have found that in contrast to intestine, gastric tissue does not yield a greater number of cells with increasing number of washing steps. We have modified the published protocol to include enough washing steps to remove the viscous gastric mucus and prevent cell damage by diluting gastric enzymes.

Reference

1. Lefrançois L, Lycke N (2001) Isolation of mouse small intestinal intraepithelial lymphocytes, Peyer's patch, and lamina propria cells. In: Coico R (ed) Current protocols in immunology. Wiley, New York, NY

Chapter 16

Delayed-Type Hypersensitivity Determination

Brian M. Gray and Kathryn A. Eaton

Abstract

Delayed-type hypersensitivity responses in the skin (in the case of mice, in the foot pad) is used to assess cell-mediated immunity (CMI) in vivo. In the case of CMI to Helicobacter infection, the mice are given an injection of cultured Helicobacter organisms into the hind footpad, and induration is measured at the site of inoculation 24 h after inoculation. Here we describe the methods for assessing delayed-type hypersensitivity in the mouse.

Key words: Foot pad, Delayed-type hypersensitivity, Mouse models

1. Introduction

1.1. Delayed-Type Hypersensitivity Determination

Persons not familiar with the techniques of foot pad injection should obtain assistance from their institutional animal care staff.

For determination of delayed-type hypersensitivity response to *H. pylori* antigen, foot pad inoculation is performed on the day before scheduled euthanasia, and the reaction is measured immedately before the mouse is euthanized.

2. Materials

1. 25 g needles and syringes.
2. Eppendorf tubes.
3. 0.45 μm filters.
4. Dial thickness gauge
5. Micro-protein assay kit

JeanMarie Houghton (ed.), *Helicobacter Species: Methods and Protocols*, Methods in Molecular Biology, vol. 921, DOI 10.1007/978-1-62703-005-2_16, © Springer Science+Business Media, LLC 2012

3. Methods

1. Thaw an aliquot of *H. pylori* sonicate and adjust the protein concentration to 33 μg/ml in sterile phosphate-buffered saline (PBS).

   To prepare *H. pylori* sonicate:

   - Grow bacteria to mid-logarithmic phase (see Chapter 13).
   - Wash three times in PBS (Gibco).
   - Quantify by hemocytometer count.
   - Resuspend to about 10^9 cfu/ml.
   - Sonicate on ice for three 30-s bursts with 30-s rests.
   - Filter sterilize (0.45 μm filter).
   - Measure protein concentration with Bio-Rad protein assay (Bio-Rad) or micro Lowry kit (Sigma).
   - Aliquot and store at –20°C.
2. Manually restrain the mouse while a second person measures paw pad thickness.
3. Measure dorsoventral thickness of both hind paws using a dial thickness gauge.
4. Using a 27 g needle, inject 10 μl of *H. pylori* sonicate into one hind pad.
5. Inject 10 μl of sterile PBS into the other hind pad. An insulin syringe works well.
6. Measure both pads again in 24 h.
7. Significant swelling occurs only in adoptively transferred infected mice, although slight swelling occasionally is present in infected C57BL/6J mice. Paws may swell to two to three times normal thickness or more, and show occasional clear discharge, but ulceration does not occur.

Chapter 17

Necropsy, Blood, Tissue Collection, and mRNA Isolation for Detection of Host Cytokine Gene Expression

Brian M. Gray and Kathryn A. Eaton

Abstract

Processing of tissue and blood must be done in a systematic and controlled fashion in order to optimize results and allow comparison of samples between experiments and between laboratories. Here we present our protocols for blood and tissue processing.

Key words: Tissue processing, Quantitative bacterial cultures, Quantitative PCR

1. Necropsy, Blood and Tissue Collection (see Note 1)

1.1. Introduction

Mice may be euthanized by isoflurane overdose or by injection of euthanasia solution. If blood is collected for serum analysis it should be done immediately after euthanasia. Many methods are used. Our laboratory has had most success with cardiopuncture, which we describe here. All personnel should be trained prior to handling mice.

2. Materials

1. TRIzol® (Invitrogen).
2. RNase ZAP.
3. cDNA kit of choice.
4. Neutral buffered formalin (NBF) (also called 10% neutral buffered formalin). This is a solution of 4% formaldehyde in phosphate buffer. It is formulated specifically for optimum fixation for histology, and can be purchased ready-to-use from Fisher.

JeanMarie Houghton (ed.), *Helicobacter Species: Methods and Protocols*, Methods in Molecular Biology, vol. 921,
DOI 10.1007/978-1-62703-005-2_17,

5. Micro-pestles or pellet pestles (Fisher).
6. TSA sheep blood agar plates treated with modified Skirrow's antibiotic supplement (see Note 4).
7. OCT freezing solution (Tissue-Tek).
8. Histology supplies (Fisher): Tissue cassettes (for formalin-fixedand frozen tissue), sponges, OCT medium for freezing, liquid nitrogen, dry ice. Forceps, surgical scissors, and scalpel with sterile blades for necropsy and tissue collection.
9. Liquid Nitrogen.
10. Micro-pestles n pellet pestles (Fisher).
11. TSA sheep blood plates treated with skirrow's antibiotic supplement (see Note 4).

3. Methods

3.1. Blood collection

1. Mice are placed in dorsal recumbency.
2. A 0.5″ 25 g needle is introduced under the skin just lateral to the xiphoid process (Fig. 5).

   The needle is held parallel to the body wall and advanced into the thoracic cavity with slight aspiration just until blood appears in the hub. Continue to apply gentle aspiration until blood flow stops. It may be necessary to rotate the needle, but avoid strong aspiration as this will collapse the vascular space, plug the needle, and/or induce hemolysis. This method yields up to 1.0 ml of whole blood.
3. Blood is sedimented in a 1.5 ml Eppendorf tube, and the serum is removed and stored at –20°C.

3.2. Necropsy—performed aseptically with sterile instruments

1. Following euthanasia, the mouse is placed in dorsal recumbency, a small incision is made in the ventral skin, and the skin is then reflected back over the head and hind legs, leaving a sterile field (Fig. 1). It is important to pull the skin back completely to prevent contamination.
2. Spleen and lymph node are collected, and placed in media on ice for rapid isolation of cells (see chapter 15).
3. The stomach is removed, incised along the greater curvature, and laid flat (Fig. 2) (see Note 2).
4. Two longitudinal strips from the greater curvature are removed and placed on absorbant paper, serosa side down, for formalin fixation or for freezing for immunohistochemistry (Fig. 3). To avoid histologic artifact, gastric mucosa must be handled gently and not squeezed or scraped with instruments.

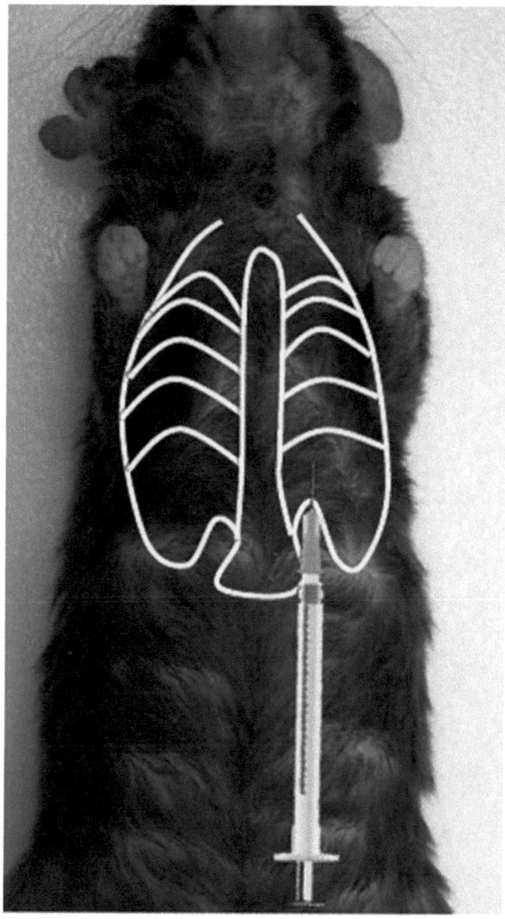

Fig. 1. Position of syringe and needle for cardiopuncture. Approximate location of sternum and ribcage is indicated.

5. For immunochemical identification of immune cells, spleen may be included as an internal control.
6. For routine histology, samples are placed in plastic cassettes and immersed in NBF overnight. Samples should be processed within 48 h of fixation to preserve antigens should immunochemical staining be anticipated.
7. Sections for freezing are placed in plastic cryostat cassettes, covered with OCT medium (Tissue-Tek), and frozen gradually over a bath of liquid nitrogen. Frozen samples are stored at −80°C.
8. The remainder of the stomach is divided for culture, DNA extraction, and/or RNA extraction, as appropriate. For detection of cytokine mRNA in tissue, samples are placed in 300 μl of TRIzol (invitrogen) in a microcentrifuge tube, and stored at −80°C until use. For DNA extraction to detect *H. pylori* colonization, samples are stored dry at −80°C until use.

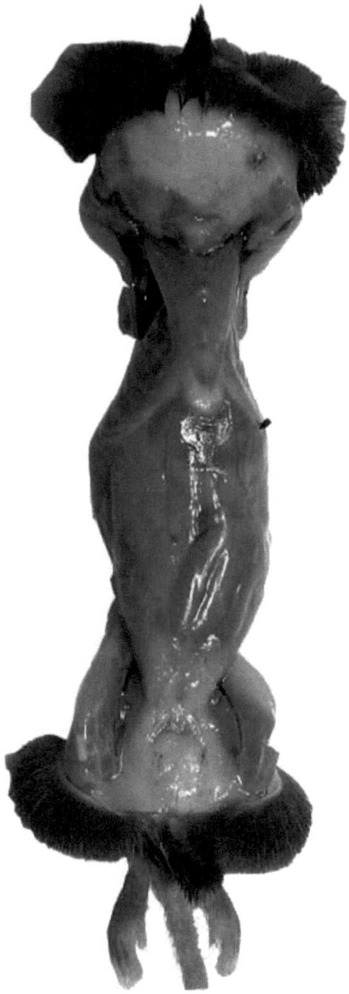

Fig. 2. Proper method of reflection of skin to ensure a sterile field for necropsy.

Generally it is possible to collect sufficient tissue for histology, culture, DNA, and RNA from a single mouse. Should additional samples be necessary, more mice must be used. The squamous mucosa and duodenum are included in samples for histology to facilitate orientation during microscopic examination. However, prior to sampling for culture and molecular analysis, ensure that all of the squamous mucosa and the duodenum have been removed and discarded.

3.3. Tissue Evaluation

Hemotoxylin- and eosin-stained formalin-fixed sections are used for scoring. Our scoring system has been published (1). It requires two longitudinal sections of gastric mucosa extending from the squamous epithelial junction to the duodenum. Sections must be well preserved and oriented so that the gastric glands are cut in longitudinal section, and twists and folds are absent. Only well-

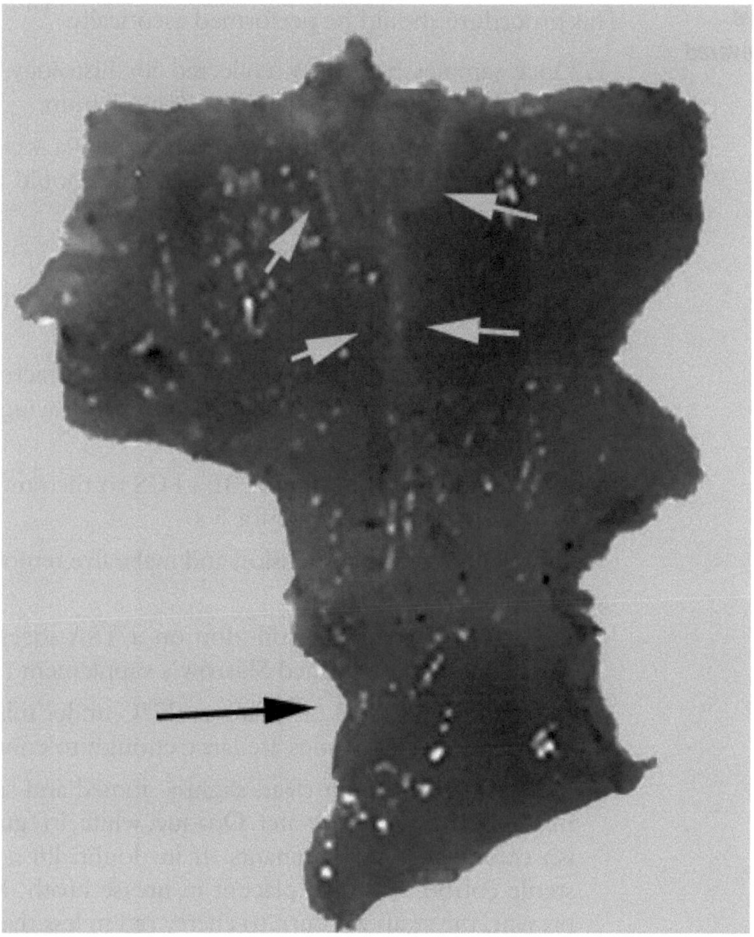

Fig. 3. Mouse stomach opened along the greater curvature. *White arrows* indicate the junction between squamous and glandular mucosa. *Black arrow* indicates the gastroduodenal junction.

oriented fields can be scored. Each 200× field of fundic mucosa is scored separately for the presence or absence of (1) neutrophilic infiltrate, (2) gastritis severe enough to displace gastric glands, and (3) mucus epithelial metaplasia (see Note 5). The percent of fields affected by each lesion is the score for that lesion. The total score is the sum of all three scores. *H. pylori* are difficult to see in HE-stained sections.

They may be visualized and scored with silver stains such as Warthin–Starry or the Steiner modification, but this rarely provides information not provided by culture and PCR. An additional stain that may be helpful is the Genta stain (2), a combination of HE, Steiner, and Alcian blue stains that differentiates mucus types as well as cell morphology and bacteria.

3.4. Quantitative Culture of Recovered Bacteria

This procedure should be performed aseptically.

1. Once samples have been collected for histology, remove and discard the squamous mucosa and duodenum.
2. Place a longitudinal strip of stomach into a weighed 1.5 ml microcentrifuge tube. The tissue sample should include both fundus and antrum.
3. Weigh and record sample mass.
4. Add 500 μl of sterile Brucella broth with 10% fetal calf serum (BB + FCS) to the tissue sample.
5. Homogenize the tissue with a disposable micro-pestle. The mucosa and submucosa will disaggregate leaving the muscularis intact.
6. Add an additional 500 μl of BB + FCS to the sample. Cap and vortex at maximum speed for 5 s.
7. Take 100 μl of the suspension and make five tenfold serial dilutions in sterile BB + FCS.
8. Plate 100 μl from each dilution on a TSA sheep blood agar plate containing modified Skirrow's supplement (above).
9. Incubate plates for 3 or 4 days at 37°C under microaerophilic conditions until colonies are large enough to count.
10. Colonies are 1–2 mm clear, slightly domed and smooth. They may appear as drops of water. Opaque, white, irregular, or greenish colonies are contaminants. If in doubt, lift a colony on a sterile cotton swab and place it in urease broth. If *H. pylori* is present, the swab will turn to cherry red in less than 5 min.
11. If growth is not visible after 4 days, re-streak the plate. For this, rub the surface of the plate with a sterile cotton swab, immediately rub the swab over the surface of a fresh blood agar plate, and incubate as above. If slow-growing colonies were present on the original plate, growth of *H. pylori* will be visible on the new plate within 3 days.

3.5. PCR for Bacterial Detection in Tissue Samples

Culture is the most accurate means of identification of *H. pylori* in tissue. However, if culture cannot be performed, *H. pylori* DNA may be detected by PCR. We use probes for *H. pylori* ureB (5′ CAGAGCGGCTGAAGAATA, 3′ TTTTAAAGCCAATCGCAC) or 16S rRNA (5′ TGGAGCCAATCTTCAAAA, 3′ GAACGTATTCACCGCAAC). These primer sequences work with *H. pylori* 26695 and SS1 strains.

1. Isolate DNA according to the Qiagen DNEasy Blood and Tissue Kit protocol (Qiagen, Valencia, CA).
2. Make PCR master mix: 1× PCR buffer, 1.5 mM $MgCl_2$, 20 μM dNTPs, 300 μM primers, and 2.5 U Taq in 50 μl total volume.

3. Add 2 μl DNA template.
4. Incubate in a thermocycler at 95°C for 5 min, followed by 36 cycles of 30 s at 95°C, 2 min and 30 s at 52°C, and 2 min at 72°C. Elongate for 4 min at 72°C.

3.6. mRNA Isolation for Detection of Host CytokineGene Expression (see Note 6)

1. Homogenize the tissue in TRIzol, using a dedicated PTFE- or Teflon-coated micro-pestle (Fisher Scientific) which is rinsed in RNAse Zap before processing each sample.
2. Isolate total mRNA according to TRIzol manufacturer's instructions, with volumes calculated for a 300 μl starting volume of TRIzol.
3. Further enrich for mRNA using Qiagen's RNEasy cleanup protocol, according to the manufacturer's instructions. mRNA can be stored in RNAse-free water at −80°C for up to 6 months without noticeable degradation.
4. Synthesize cDNA with the SABiosciences RT^2 First Strand kit (or equivalent) to generate cDNA libraries prior to performing qRT-PCR. Store cDNA at −20°C or colder for up to 1 year.

4. Notes

1. *Experimental interval.* The duration of the experiment will depend on the experimental design and information to be collected. *H. pylori* colonization becomes established within 1 week of inoculation, and remains stable until inflammation develops (3–5). Therefore, a 1-week duration is sufficient for experiments to determine colonization efficiency. Gastritis in C57BL/6J mice develops slowly over several months. Usually it is detectable 8–12 weeks after infection, and becomes moderate to severe by 6 months. In adoptively transferred mice, gastritis is detected by 2–4 weeks after transfer and becomes severe by 8 weeks. Thus, experiments to evaluate aspects of disease are longer, generally from 8 weeks for recipient mice to 12 months for C57BL/6J mice. Severe gastric metaplasia, used as a model for preneoplasia (see Note 5), develops late in the course of disease (8–12 weeks in recipient mice, 12–18 months in C57BL/6J mice).
2. *Mouse gastric anatomy.* Mouse stomachs differ from human stomachs in several respects. First, mice have a large non-glandular part of the stomach, just distal to the esophagus, that is lined by stratified squamous epithelium similar to that lining the esophagus. The junction between glandular and non-glan-

dular (or squamous) mucosa is grossly visible (see Fig. 4). *H. pylori* colonizes the glandular region only, but the squamous mucosa is colonized by many bacterial species, including lactobacilli that may interfere with recovery of *H. pylori* (Fig. 5). Thus, removal of the entire squamous mucosa prior to culture is indicated. Second, mice have few actual cardiac glands. Most of the glandular stomach consists of fundic glands proximally and along the fundus, and pyloric glands distally and along the lesser curvature. Very few fundic glands are present on the lesser curvature. For this reason, sampling of all gastric gland types requires sampling of the greater curvature. The gastric lymph node is small and is located at the junction of esophagus and stomach. With practice, it can be located and collected, but it may contain very few recoverable lymphocytes.

3. *Histopathologic scoring*. Several considerations are important regarding histopathologic scoring. Three principles to keep in mind are the following: (a) histopathologic interpretation is a medical specialty requiring considerable training and experience; (b) mouse tissues and mouse gastric response to injury differ from those of humans; and (c) histologic interpretation necessarily incorporates a certain amount of subjective analysis. We have developed the scoring system used in our laboratory to facilitate reproducible quantification of gastritis in mice by observers of various levels of knowledge and experience. Features of this system are the following: (a) it can be learned relatively easily by observers without extensive pathology experience; (b) it is quantifiable and minimizes observer bias; (c) it is specific to lesions characteristic of mouse models of helicobacter gastritis; and (d) it is reproducible and reflects independent measures of disease severity (1). Other commonly used methods of scoring include subjective semiquantification of lesions (6, 7) and scoring based on the Sydney system, designed for evaluation of human tissues (8). The former method is highly dependent on access to a veterinary pathologist who is trained and experienced in evaluating mouse tissues, and requires that all tissues be scored by the same person to minimize individual differences in interpretation. The latter method scores lesions that are not generally present in mouse tissues and fails to score lesions that are commonly present. Whichever method of scoring is used, it is necessary to consult with an experienced veterinary mouse pathologist if not to score the slides, and then to train the observers to recognize and score the histologic lesions.
4. *Antibiotic-treated agar plates*. As an alternative to including antibiotics in the plates during pouring, antibiotics may be added to prepurchased plates. Pipet the appropriate volume on the plate, spread the solution over the entire surface, and allow the plate to dry before use.

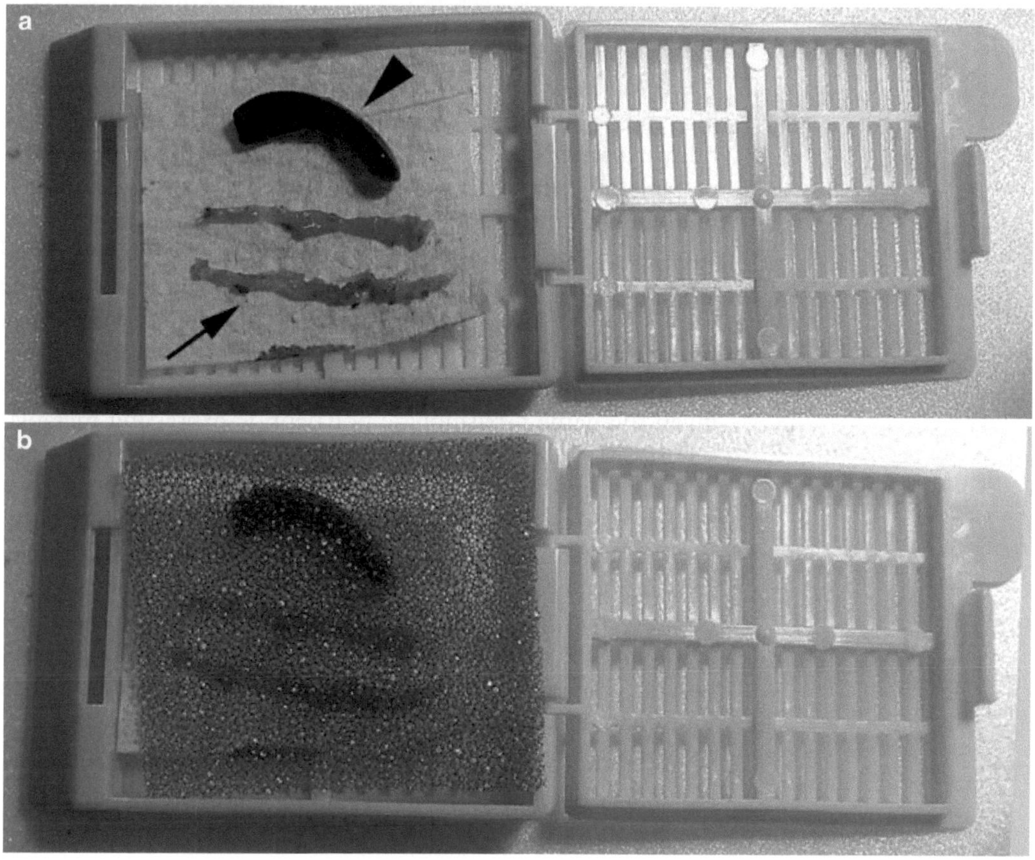

Fig. 4. (**a**) Two longitudinal strips of greater curvature of the stomach (*arrow*) are placed serosa-side down on absorbant paper in a histology cassette prior to emersion in 10% NBF. *Arrowhead* indicates placement of the spleen. (**b**) Placement of sponge to secure tissue in place prior to closing the cassette.

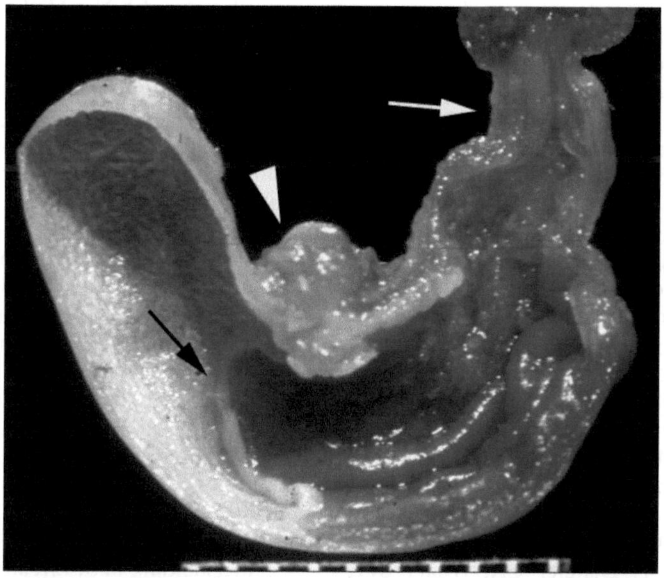

Fig. 5. Gross appearance of a normal mouse stomach fixed in situ. Junctions between the glandular and non-glandular mucosa (*black arrow*) and between the pylorus and duodenum (*arrowhead*) are clearly visible. The white arrow indicates the approximate position of the gastric lymph node.

5. *Epithelial metaplasia/SPEM*. Considerable confusion regarding gastric epithelial metaplasia exists in the literature. In human patients, *H. pylori* may be associated with intestinal metaplasia, characterized by replacement of gastric glandular mucosa with intestinal epithelium containing goblet cells and sometimes developing villi (9). Chronic infection is associated with gastric epithelial atrophy characterized by loss of gastric glands and replacement by fibrosis (9). These lesions are not present in mice with gastric helicobacter. Mice (and other rodents) develop a different kind of metaplasia that has been given the name "spasmolytic polypeptide-expressing metaplasia (SPEM)" (10, 11). SPEM is characterized by loss of parietal and chief cells within gastric glands and replacement by a mucus-type epithelium that expresses mucin types more common in intestinal epithelium. Intestinal epithelium and goblet cells are not present, and atrophy is limited to atrophy of parietal and chief cells. Gastric glands remain intact. The characteristics of SPEM have been reviewed (11). SPEM has been described in human biopsies, but appears less common than intestinal metaplasia in association with *H. pylori* infection.
6. Some laboratories have reported detection of gastric cytokines by ELISA (12, 13), but we have not found this to be reproducible in mice.

References

1. Eaton KA, Danon SJ, Krakowka S, Weisbrode SE (2007) A reproducible scoring system for quantification of histologic lesions of inflammatory disease in mouse gastric epithelium. Comp Med 57:57–65
2. Genta RM, Robason GO, Graham DY (1994) Simultaneous visualization of *Helicobacter pylori* and gastric morphology: a new stain. Hum Pathol 25:221–226
3. Eaton KA, Ringler SR, Danon SJ (1999) Murine splenocytes induce severe gastritis and delayed-type hypersensitivity and suppress bacterial colonization in *Helicobacter pylori*-infected SCID mice. Infect Immun 67:4594–4602
4. Eaton KA, Mefford M, Thevenot T (2001) The role of T cell subsets and cytokines in the pathogenesis of *Helicobacter pylori* gastritis in mice. J Immunol 166:7456–7461
5. Eaton KA, Mefford ME (2001) Cure of *Helicobacter pylori* infection and resolution of gastritis by adoptive transfer of splenocytes in mice. Infect Immun 69:1025–1031
6. Fox JG, Beck P, Dangler CA, Whary MT, Wang TC, Shi HN, Nagler-Anderson C (2000) Concurrent enteric helminth infection modulates inflammation and gastric immune responses and reduces helicobacter-induced gastric atrophy. Nat Med 6:536–542
7. Ghiara P, Marchetti M, Blaser MJ, Tummuru MKR, Cover TL, Segal ED, Tompkins LS, Rappuoli R (1995) Role of the *Helicobacter pylori* virulence factors vacuolating cytotoxin CagA, and urease in a mouse model of disease. Infect Immun 63:4154–4160
8. Stolte M, Meining A (2001) The updated Sydney system: classification and grading of gastritis as the basis of diagnosis and treatment. Can J Gastroenterol 15:591–598
9. Correa P, Houghton J (2007) Carcinogenesis of *Helicobacter pylori*. Gastroenterology 133: 659–672
10. Rogers AB, Fox JG (2004) Inflammation and Cancer. I. Rodent models of infectious gastrointestinal and liver cancer. Am J Physiol Gastrointest Liver Physiol 286: G361–366
11. Weis VG, Goldenring JR (2009) Current understanding of SPEM and its standing in the preneoplastic process. Gastric Cancer 12: 189–197

12. Shi Y, Liu XF, Zhuang Y, Zhang JY, Liu T, Yin Z, Wu C, Mao XH, Jia KR, Wang FJ, Guo H, Flavell RA, Zhao Z, Liu KY, Xiao B, Guo Y, Zhang WJ, Zhou WY, Guo G, Zou QM (2010) *Helicobacter pylori*-induced Th17 responses modulate Th1 cell responses, benefit bacterial growth, and contribute to pathology in mice. J Immunol 184:5121–5129

13. Fox JG, Wang TC, Rogers AB, Poutahidis T, Ge Z, Taylor N, Dangler CA, Israel DA, Krishna U, Gaus K, Peek RM Jr (2003) Host and microbial constituents influence *Helicobacter pylori*-induced cancer in a murine model of hypergastrinemia. Gastroenterology 124:1879–1890

Chapter 18

Animal Models of *Helicobacter*-Induced Disease: Methods to Successfully Infect the Mouse

Nancy S. Taylor and James G. Fox

Abstract

Animal models of microbial diseases in humans are an essential component for determining fulfillment of Koch's postulates and determining how the organism causes disease, host response(s), disease prevention, and treatment. In the case of *Helicobacter pylori*, establishing an animal model to fulfill Koch's postulates initially proved so challenging that out of frustration a human volunteer undertook an experiment to become infected with *H. pylori* and to monitor disease progression in order to determine if it did cause gastritis. For the discovery of the organism and his fulfillment of Koch's postulates he and a colleague were awarded the Nobel Prize in Medicine. After *H. pylori* was established as a gastric pathogen, it took several years before a model was developed in mice, opening the study of the organism and its pathogenicity to the general scientific community. However, while the model is widely utilized, there are a number of difficulties that can arise and need to be overcome. The purpose of this chapter is to raise awareness regarding the problems, and to offer reliable protocols for successfully establishing the *H. pylori* mouse model.

Key words: *H. pylori* mouse model, Mouse orogastric inoculation, *H. pylori* culture, *H. pylori*-specific PCR, Helicobacter genus PCR, Microaerobic, *H. pylori* selective media

1. Introduction

With the discovery of *Helicobacter pylori* and its association with gastric disease in humans (1), the search for an animal model to study the organism and its pathogenesis became of great importance. Even more interest in an animal model was generated when *H. pylori* was classified as a Class 1 carcinogen by the World Health Organization (2). A mouse model was the most logical approach from a cost and housing consideration. However, it soon became apparent that it is very difficult to persistently colonize immunocompetent mice with *H. pylori* (3–5). Short-term colonization

JeanMarie Houghton (ed.), *Helicobacter Species: Methods and Protocols*, Methods in Molecular Biology, vol. 921, DOI 10.1007/978-1-62703-005-2_18,

(2–4 weeks at most) was reported by a few laboratories, but this was not promising with respect to development and study of chronic gastritis and peptic ulcer. It also made treatment and immunization studies difficult. Other animal models described were gnotobiotic pigs (6), conventional piglets (7), primates (8), cats (9), germ-free mice (10), nude mice (11), gnotobiotic beagle dogs (12), and Mongolian gerbils (13) which are the only wild-type rodent species that develop gastric carcinoma following *H. pylori* infection without other tumor-promoting factors (14).

A promising mouse model was reported by Lee et al. (15) using *Helicobacter felis* which was originally isolated from the stomach of cats (16). A close relative of *H. pylori*, this organism colonizes mouse stomachs long term and can produce low-grade B-cell gastric lymphomas indistinguishable from those in infected humans (17). Because of the ease of colonization and similarity to human disease, this model became widely used. The ferret was found to be naturally infected with *H. mustelae* which causes chronic gastritis (18) and was also initially explored as a potential model for human *H. pylori* infection (19). However, while these models generated data that was highly relevant to *H. pylori* in humans, they used organisms that are similar to *H. pylori*, but they are not *H. pylori*. It was not until Lee et al. isolated the Sydney strain of *H. pylori* and showed it could colonize mice long term that the mouse model using *H. pylori* became viable for a wide range of research laboratories (20). Using this mouse-adapted strain, good levels of colonization were maintained in conventional C57BL/6 and Balbc mice. After 8 months, chronic active gastritis developed which progressed to severe atrophy. This laboratory also assessed the number of organisms needed to colonize mice and also looked at the susceptibility of different strains of mice to colonization (21). They found that C57BL/6> Balbc> Swiss Webster mice were colonized with *H. pylori* in that order with C57BL/6 mice exhibiting the best colonization. In 2004, a second mouse-adapted *H. pylori* isolate with virulence factors differing from *H. pylori* Sydney was described by Thompson et al. to serve as a model for determining the importance of various virulence factors in the progression of disease (22). Another strain of *H. pylori* reported to cause disease in the mouse model is the B128 strain (23, 24). Table 1 lists some of the conventional strains of mice susceptible to *H. pylori* and *H. felis* colonization. Rogers and Houghton have written a review describing the mouse model in enterohepatic helicobacter-induced carcinogenesis, including *H. pylori*, and include a description of genetically engineered mice susceptible to development of cancer (25).

The *H. pylori* Sydney strain remains the organism of choice for use as a model of human disease. However, this model can be very difficult to reproduce in the laboratory due to a number of factors including culture viability, inoculum, frequency of dosing, repeated

Table 1
Mouse susceptibility to *H. pylori* and *H. felis* challenge

Mouse strain	Colonization by *H. pylori* SS1	Pathology	Colonization by *H. felis*	Pathology
C57BL/6	Yes	Chronic active gastritis progressing to severe hyperplasia and atrophy	Yes	Chronic active gastritis progressing to severe hyperplasia and atrophy
Balbc	Yes	Chronic active gastritis progressing to severe atrophy	Yes	Some gastric atrophy, less apoptosis, mild antral gastritis, MALToma
SJL	Yes, but low numbers	Mild gastritis	Yes	Corpus gastritis
C3H/He	Yes, but low numbers	Moderate to severe chronic active gastritis	Yes	Corpus gastritis
DBA/2	Yes, but low numbers		Yes	Corpus gastritis
CBA	Yes, but low numbers	Mild gastritis	Yes	Mild antral gastritis
Swiss Webster	Yes, but low numbers	Mild gastritis	Yes	Moderate self-limiting gastritis

passage of the organism, and mouse strain chosen. The protocols detailed below address these problems and provide guidance aimed at repeated successful colonization.

2. Materials Needed

2.1. Media

1. Autoclave.
2. Pipetman and pipet tips.
3. 1.5 ml Cryovials.
4. Brucella broth/agar.
5. Trypticase soy agar plates.
6. Blood Agar Base.
7. Bacitracin.
8. Amphotericin B.
9. Vancomycin.
10. Polymyxin B.

11. Naladixic Acid.
12. Sterile standard petri dishes (100 × 20 mm).
13. Defibrinated Horse blood.
14. Brucella broth with 20% glycerol.
15. Urea agar slants.
16. Fetal bovine serum.
17. Freeze media: 80 ml of double-distilled water to the amount of brucella broth powder needed for 100 ml total volume, and then adding 20 ml of glycerol. This is referred to as freezing medium; freezing media.

2.2. Culture

1. 37°C Incubator with rotating platform.
2. BD GasPak system vented jars (Becton Dickenson, cat no. 260627).
3. Anaerobic gas mix (80:10:10/N_2:CO_2:H_2).
4. Tubing (VWR red vacuum tubing, cat. No. 62995-059).
5. Clamps.
6. T fittings.
7. Vacuum pump (capable of vacuum down to 20 in. of mercury).
8. 37°C Incubator.
9. Sterile swabs.
10. Spectrophotometer (visible capability).
11. Phase contrast microscope.
12. Light microscope.
13. Gram's stain reagents.
14. 250 ml Flasks with screw caps or sliding metal caps.
15. Biosafety hood.

2.3. Mouse Inoculation

1. 1 cc Syringes.
2. 24 gauge × 1 in. stainless steel feeding tubes.

2.4. PCR

1. Thermocycler.
2. Bovine serum albumin.
3. dNTPs.
4. Taq polymerase.
5. Primers: C97 (5′-GCT ATG ACGGGT ATCC) and C05 (5′-ACT TCA CCC CAG TCG CTG). Hp1: AAA GCT TTT AGG GGT GTT AGG GGT TT and Hp2:AAG CTT ACT TTC TAA CAC TAA CGC (26).
6. PCR Buffer.
7. Sterile distilled water.

8. Eppendorf tubes.
9. Agarose.
10. Gel box.
11. Electrophoresis power supply.
12. Ethidium bromide.
    Dilute a sample of the stock solution to 0.5 μg/ml with water. Protect from light.
13. Documentation center for taking photos of gels.
14. 1 kb Plus DNA Ladder.
15. TAE Buffer.

3. Methods

3.1. Microbial Stock Solutions

Excessive passage of the isolates is known to affect colonization and virulence, so it is necessary to freeze several vials of inoculum to last for about 6 months. The number of stock vials that you keep will depend on the number of experiments you anticipate.

1. One dram vials, or equivalent, are filled with 1 ml of brucella broth containing 20% glycerol.
2. The vials are them autoclaved and stored at room temperature.
3. Stock is prepared from *H. pylori* grown on agar plates (trypticase soy blood agar) for 2–3 days. Do not use cultures that are older as they will yield poor, if any, growth on plating.
4. Growth is harvested from the plates using a sterile swab dipped in freezing medium and the swab swirled in a sterile, cool vial. The medium in the vial should then be lightly turbid. Several vials can be prepared from one plate.
5. These vials are then frozen at −80°C. Freezing at −20°C is not recommended, nor is refreezing thawed vials of stock as the organism will not be viable (see Note 1).

3.2. Growth of H. pylori for Mouse Inoculation

All cultures are incubated in a microaerobic environment. *H. pylori* can be grown on blood agar plates or in broth for mouse inoculation. (see Note 2).

3.2.1. Selective Media

1. The antibiotics in Table 2 are added to autoclaved and cooled Blood Agar Base (Sigma Chemical Company) along with 5% sterile horse blood and after gentle but thorough mixing to avoid bubbles, poured into standard sized petri dishes.
2. This media can be kept at 4°C for up to 2 months if stored in plastic containers to keep the plates moist.
3. Always do quality assurance using a known *H. pylori* to ensure good growth.

Table 2
Antibiotics for selective media preparation

Antibiotic	Final concentration	Sigma catalogue number
Amphotericin B, solubilized	50 μg/ml	A 9528 50 mg (~45% Amph. B
Vancomycin HCl	100 μg/ml	V 2002 1 g
Polymyxin B sulfate	3.3 μg/ml	P 1004 5,000,000 units (~7,870 U/mg)
Bacitracin	200 μg/ml	B 0125 50,000 units
Naladixic acid, Na salt	10.7 μg/ml	N 4382 1 g

3.2.2. Blood Agar Plates

1. Either commercially bought or in-house prepared plates may be used but the plates should be fresh and moist.
2. Trypticase soy agar, brain–heart infusion agar, brucella agar, or Blood Agar Base can be used but all must be supplemented with 5% whole blood or fetal calf serum.
3. Serum can be used in place of whole blood as red blood cells are not necessary for good growth. However, the use of blood agar makes the *H. pylori* colonies more easily visible since they have a watery appearance when confluent that often looks like "no growth" to the novice. Either horse blood or sheep blood can be used. Blood may be stored frozen and thawed for making plates if in-house plates are prepared. Lysis of the blood does not affect the growth of *H. pylori* though, again, colony visibility can be an issue on translucent plates, especially if quantitative cultures are performed. An added advantage of using agar plates is that you can visually detect contamination.
4. Plates should be inoculated using a sterile swab dipped into a thawed stock vial to yield a heavy, confluent growth after 2–3 days of microaerobic incubation.
5. Approximately 2 ml of inoculum can be obtained from each plate.
6. Growth is harvested using a sterile swab moistened in brucella broth. After swiping the plate several times, the organisms are harvested by swirling the swab in brucella broth. The plate can be swiped a second time to get as much of the growth as possible.
7. At the same time as preparing this inoculum, inoculate new plates with the swab and incubate to use for your second inoculation of the mice in a day or two (see Notes 3 and 4).
8. When all plates have been harvested, the motility and morphology of the organisms should be observed using a phase contrast microscope. While there will be more coccoid forms (round balls instead of gently curving spirals) and less motility in

cultures from agar plates as compared to broth, it is important that the majority of organisms (roughly 80% or more) have good morphology and are motile. If the culture is too old, coccoid forms will predominate and these will not colonize mice.

9. The optical density is then measured at a wavelength of 660 nm. The optical density should be between 0.8 and 1.4 to ensure enough viable organisms for good colonization. As mice receive 3 doses of organism, repeat this procedure a total of three times in order to prepare fresh inoculum for each of the 3 doses. At the end of 3rd harvest session, all the remaining cultures are discarded and the process from thawing stock to preparation of inoculum started totally anew for subsequent experiments.

3.2.3. Broth Culture

1. Broth culture is more difficult because contamination is not readily apparent and the likely hood for contamination is increased by the manipulations of the flask. However, morphology and motility are exceptional in a young broth culture. Five hundred milliliter flasks with screw caps work well, though flasks with metal sliding caps can also be used.
2. Each flask is filled with 150 ml of brucella broth and then autoclaved. On cooling, 7.5 ml (5%) of fetal calf serum is added.
3. Flasks are inoculated with an *H. pylori* suspension prepared from freshly grown agar plates inoculated from stock as described above. It is not recommended to inoculate the flask directly from frozen stock because the lag time for the organism to grow well can take several days. Each flask should be inoculated with 1 ml of suspension.
4. Sliding metal caps allow for good gas exchange with the culture, but be careful that screw caps are left loose.
5. The flasks are then placed in jars that can be vented. Two-sided tape on the bottom of the jar prevents the flask from sliding during shaking.
6. After venting the jars to obtain a microaerobic environment, the jars are incubated at 37°C overnight with shaking on an orbital shaker at 100 rpm.
7. When turbidity is observed, and this may be on day 2, using phase microscopy, assess the morphology and motility. There should be minimal (<10%) coccoid forms and good motility.
8. New flasks are inoculated with 1 ml of the culture from the turbid flasks and then incubated as before for the next mouse inoculation (see Note 1).
9. The broth is then centrifuged at 8,450×g for 20 min, and the supernatant discarded.
10. The pellet is suspended in 10 ml of Brucella broth and the optical density at 660 nm determined. Adjust the optical density to between 0.8 and 1.4 for your inoculum using brucella broth.

3.2.4. Obtaining a Microaerobic Environment

1. All agar plates and flasks are placed in jars with lids that enable a good seal and a vent allowing for evacuation of the jar and filling with gas mixture. An example of this type of jar is the BBL vented jars used with the Gas-Pak system. Use of an anaerobic mix (80:10:10/N_2:CO_2:H_2 or 90:5:5/N_2:CO_2:H_2) in a vented jar system works best.
2. Use a vacuum pump connected by tubing through a T fitting to tubing to a gas gauge. The tubing from the third port of the T is connected to another T fitting that is connected to the gas tank and the third port is for attaching to the jar.
3. A clamp is placed between the vacuum pump and the first T fitting (Fig. 1). Using tubing on the vent of the jar lid, connect the vent on the jar to the unused port of the second T fitting and make sure that the clamp is not engaged.
4. Turning on the vacuum pump, evacuate the jar down to 20 in. of mercury as indicated by the gas gauge.
5. Use the clamp to close the connection to the pump. Turn off the vacuum pump. The vacuum in the jar should hold at 20 in. If it does not, then the jar is not sealed and you need to use another, or determine why it is leaking. Once you have determined that the jar is sealed, slowly fill the jar with the gas mix

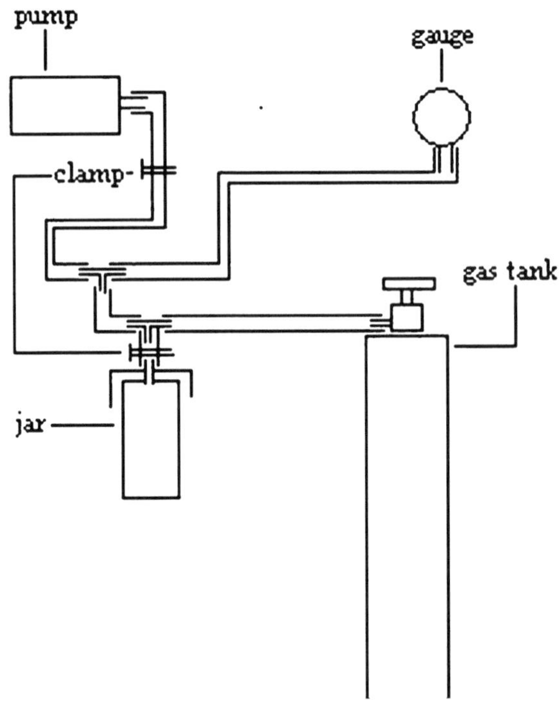

Fig. 1. Venting system for microaerobic culture conditions.

until the gas gauge reads 0. If you fill too rapidly, you will blow fungi, etc. into the flasks/plates.

6. Clamp off the tubing on the jar vent and incubate as directed above (see Note 5).

3.3. Inoculation of the Mice

Only personnel who are appropriately trained and approved by the institution's Committee on Animal Care should orally gavage mice. This is a very exacting procedure and esophageal damage and pulmonary aspiration and/or tracheal damage can result in the immediate or prolonged death of the mouse.

1. For *H. pylori* inoculation, each mouse receives 0.2 cc of inoculum intragastrically using a 24 gauge × 1 in. straight stainless steel feeding tube attached to a 1 cc syringe. This is done without anesthesia. This results in an inoculum of roughly 2×10^8 organisms. The mice will receive a total of 3 doses, 1 per day and each separated by 1–2 days.

3.4. Verification of Colonization

1. Two to 3 weeks post inoculation, if verification of colonization is required one or two mice should be euthanized and the gastric tissue assessed by culture and/or PCR for *H. pylori* (see Note 3).
2. Intestinal shedding of other enterohepatic *Helicobacter* spp. allows PCR of fecal samples to determine colonization, thus sparing the need to euthanize an animal.
3. Shedding of *H. pylori* in feces is not a reliable indicator for colonization. If PCR is positive for *H. pylori*, then colonization is confirmed. A negative PCR does not mean that mice are not colonized.

3.4.1. Culture of H. pylori from Gastric Tissue (see Note 4)

1. A small piece (approximately 1 or 2 mm square) of antrum or body tissue should be collected into a sterile vial (similar to a 1 dram vial) containing 0.5–1 cc of freezing medium and kept on ice until processed.
2. The sample can be frozen at −80°C, or processed immediately.
3. To process, use a sterile glass tissue grinder (3 ml capacity), pour the tissue and broth into the tissue grinder tube, and homogenize with the glass pestle until the suspension is homogeneous. Disposable tissue grinders are not recommended for gastric tissue as the tissue is too difficult to homogenize.
4. Plate 0.1 cc of the homogenized tissue onto a small section of selective media (below) and streak to obtain isolated colonies.

5. Incubate the plate microaerobically for 3–5 days, checking at 3 days and 5 days if necessary. Individual colonies are small (1 mm or less) and water droplet in appearance (see Note 5).
6. To ensure that these are *H. pylori*, a rapid urease test (below) should be performed and a Gram's stain examined for typical morphology.

3.4.2. PCR for H. pylori

1. Tissue for PCR should not be placed into brucella broth with glycerol as the glycerol interferes with the PCR. Collect these samples in an empty vial.
2. These can be processed immediately or frozen at –20°C. This temperature is sufficient as the DNA is to be analyzed and viability is not an issue.
3. Tissue is homogenized in 200 μl of sterile PBS using sterile disposable plastic tissue grinders. These are acceptable for PCR samples because they have not been exposed to helicobacter-infected tissues previously, and the resultant homogenate, though not homogeneous, will be further digested with the reagents in the kit.
4. Extract using the High Pure PCR Template Preparation Kit (Roche Molecular Biochemicals, Indianapolis, Ind.) using the tissue extraction protocol.
5. *Helicobacter* genus-specific primers C97 and C05 are used to amplify a 1.2-kb PCR product from the 16SrRNA gene. These primers are excellent for detecting *Helicobacter* species and will be reliable only if mice are not colonized with other *Helicobacter spp.*
6. Ten microliters of the DNA preparation is used for a PCR 100 μl reaction mix.
7. The PCR mixture contains 10 μl Taq polymerase buffer, 0.5 mM each of the two primers, 200 mM each deoxynucleotide, and 200 mg of bovine serum albumin per ml and 0.53 μl Taq polymerase.
8. Amplification conditions are as follows: initial denaturation at 94°C for 5 min, then 94°C for 1 min, annealing at 58°C for 2 min, and elongation at 72°C for 3 min for a total of 35 cycles completed before a final elongation step at 72°C for 8 min.
9. A 15 μl aliquot of the PCR product is electrophoresed through a 1% agarose gel separation matrix prior to ethidium bromide staining and viewing under a UV light.
10. If other *Helicobacter* spp. are present in the mice, *H. pylori*-specific primers HP1, Hp2 can be used to verify colonization. They are used in the same proportions and cycling conditions described above for the genus primers.

3.4.3. Rapid Urease Test

Using a straight wire inoculating needle, pick an individual colony and stab the needle into a urea agar slant.

1. For a positive test, after no more than 2–3 min, the stab should be turning a bright pink.

4. Notes

1. Replace frozen stock at least every 6 months with a fresh isolate from an experimental animal with good gross and histologic lesions. This is important because *H. pylori* has been known to lose its ability to colonize on prolonged storage and also on multiple passages on laboratory media.
2. It may seem like a lot of work to prepare your own media, and *H. pylori* will grow on other selective media, but many other gastric organisms compete, especially lactobacilli, making it difficult to identify the *H. pylori*. Almost nothing else will grow on the selective media, making life a lot easier in the long run.
3. Always prepare your next inoculation plates/flasks first before you do any other manipulation of the culture in order to minimize the risk of contamination.
4. Always use Gram's stain broth cultures and agar plate cultures to ensure a pure culture.
5. A microaerobic environment can also be achieved using BBL-type jars and Campy Paks (or similar) or anaerobic paks without catalyst, but it is not known if all *H. pylori* isolates will grow well using this system. Many *Helicobacter* spp. grow best in the presence of hydrogen. Using the GasPak system does, however, negate the need for gas tanks, vacuum pumps, and vented jars.

References

1. Marshall BM, Armstrong JA, McGechie DB, Glancy RJ (1885) Attempt to fulfil Koch's postulates for pyloric campylobacter. Med J Aust 142:436–439
2. IARC (1994) Schistosomes, liver flukes, and *Helicobacter pylori*. IARC Working Group on Evaluation of carcinogenic risks to humans. Lyon, 7–14 June 1994. IARC Monogr Eval Carcinogen Risks Hum. 61:1.
3. Cantorna MT, Balish E (1990) Inability of human clinical strains of *Helicobacter pylori* to colonize the alimentary tract of germ-free rodents. Can J Microbiol 36:237–241
4. Marchetti M, Arico B, Burroni D, Figura N, Rappouli R, Ghiara P (1995) Development of a mouse model of *Helicobacter pylori* infection that mimics human disease. Science 267:1655–1658
5. Lachman LB, Ozpolat B, Rao XM, Graham DY, Osato M (1997) Development of a murine model of *Helicobacter pylori* infection. Helicobacter 2:78–81
6. Krakowka S, Eaton KA, Rings DM, Morgan DR (1991) Gastritis induced by *Helicobacter pylori* in gnotobiotic piglets. Rev Infect Dis 8:S681–S685

7. Poutahidis T, Tsangaris T, Kanakoudis G, Vlemmas I, Iliadis N, Sofianou D (2001) *Helicobacter pylori*-induced gastritis in experimentally infected conventional piglets. Vet Pathol 38:667–678
8. Dubois A, Berg D, Incecik E, Fiala N, Heman Ackah L, Yang M, Wirth H, Perez-Perez GI, Blaser MJ (1999) Host specificity of *Helicobacter pylori* strains and host responses in experimentally challenged nonhuman primates. Gastroenterology 116:90–96
9. Fox JG, Perkins S, Yan L, Shen Z, Attardo L, Pappo J (1996) Local immune response in *Helicobacter pylori*-infected cats and identification of *H. pylori* in saliva, gastric fluid and faeces. Immunol 88:400–406
10. Karita M, Li Q, Cantero D, Okita K (1994) Establishment of a small animal model for human *Helicobacter pylori* infection in the germ free mouse. Am J Gastroenterol 89:208–213
11. Karita M, Li Q, Okita K (1993) Evaluation of new triple therapy for eradication of *H. pylori* infection in nude mouse model. Am J Gastroenterol 88:1366–1372
12. Radin JM, Eaton KA, Krakowa S, Lee A, Otto G, Fox JG (1990) Helicobacter pylori infection in gnotobiotic beagle dogs. Infect Immun 60:2606–2612
13. Yokita K, Kurebayashi Y, Takayama Y, Hayashi S, Isogai H, Isogai E, Imai K, Yabana T, Yachi A, Oguma K (1991) Colonization of Helicobacter pylori in the gastric tissue of Mongolian gerbils. Microbiol Immunol 35:475–480
14. Hirayama F, Takagi S, Kusuhara H, Iwao E, Yokoyama Y, Ikeda Y (1996) Induction of gastric ulcer and intestinal metaplasia in Mongolian gerbils infected with *H. pylori*. J Gastroenterol 31:755–757
15. Lee A, Fox JG, Otto G, Murphy J (1990) A small animal model of human Helicobacter pylori chronic gastritis. Gastroenterology 99:1315–1325
16. Paster BJ, Lee A, Fox JG, Dewhirst FE, Tordoff LA, Fraser GJ, O'Rourke JL, Ferrero R (1991) Phylogeny of *Helicobacter felis* sp. nov., *Helicobacter mustelae*, and related bacteria. Int J Syst Bacteriol 41:31–38
17. Enno A, O'Rourke JL, Howlett CR, Jack A, Dixon MF, Lee A (1995) MALToma-like lesions in the gastric mucosa after long term infection with Helicobacter felis. Am J Pathol 147:217–222
18. Fox JG, Cabot EB, Taylor NS, Laraway R (1988) Gastric colonization by *Campylobacter pylori* subsp. *mustelae* in ferrets. Infect Immun 56:2994–2996
19. Fox JG, Correa P, Taylor NS, Lee A, Otto G, Murphy JC, Rose R (1990) *Helicobacter mustelae*-associated gastritis in ferrets. An animal model of Helicobacter pylori gastritis in humans. Gastroenterology 99:352–361
20. Lee A, O'Rourke J, De Ungria J, Robertson B, Daskalopoulos G, Dixon MF (1997) A standardized mouse model of *Helicobacter pylori* infection: introducing the Sydney strain. Gastroenterology 112:1386–1397
21. Ferrero RL, Thiberge J-M, Huerre M, Labigne A (1998) Immune responses of specific-pathogen free mice to chronic *Helicobacter pylori* (strain SS1) infection. Infect Immun 66:1349–1355
22. Thompson LJ, Danon SJ, Wilson JE, O'Rourke JL, Salama NR, Falkow S, Mitchell H, Lee A (2004) Chronic *Helicobacter pylori* infection with the Sydney strain 1 and a newly identified mouse-adapted strain (Sydney strain 2000) in C57BL/6 and Balbc mice. Infect Immun 72:4668–4679
23. Israel DA, Salama N, Arnold CN, Moss SF, Ando T, Wirth HP, Tham KT, Camorlinga M, Blaser MJ, Falkow S, Peek RM Jr (2001) *Helicobacter pylori* strain-specific differences in genetic content, identified by microarray, influence host inflammatory responses. J Clin Invest 107:611–620
24. Fox JG, Wang TC, Rogers AB, Poutahidis T, Ge Z, Taylor NS, Dangler CA, Israel DA, Krishna U, Gaus K, Peek RM (2003) Host and microbial constituents influence *Helicobacter pylori*-induced cancer in a model of hypergastnemia. Gastroenterology 124:1879–1890
25. Rogers A, Houghton J (2009) Helicobacter-based mouse models of digestive system carcinogenesis. In: Kozlov SV (ed) Inflammation and cancer: methods in molecular biology, vol 511. pp 267–295
26. Bickley J, Owen RJ, Fraser AG, Pounder RE (1993) Evaluation of the polymerase chain reaction for detecting the urease C gene of *Helicobacter pylori* in gastric biopsy samples and dental plaque. J Med Mocrobiol 39:338–344

Chapter 19

Verifying and Quantifying *Helicobacter pylori* Infection Status of Research Mice

Mark T. Whary, Zhongming Ge, and James G. Fox

Abstract

Mice used to model helicobacter gastritis should be screened by PCR prior to experimental dosing to confirm the absence of enterohepatic *Helicobacter* species (EHS) that colonize the cecum and colon of mice. Natural infections with EHS are common and impact of concurrent EHS infection on *Helicobacter pylori*-induced gastric pathology has been demonstrated.

PCR of DNA isolated from gastric tissue is the most sensitive and efficient technique to confirm the *H. pylori* infection status of research mice after experimental dosing. To determine the level of colonization, quantitative PCR to estimate the equivalent colony-forming units of *H. pylori* per μg of mouse DNA is less labor-intensive than limiting dilution culture methods. Culture recovery of *H. pylori* is a less sensitive technique due to its fastidious in vitro culture requirements; however, recovery of viable organisms confirms persistent colonization and allows for further molecular characterization of wild-type or mutant *H. pylori* strains. ELISA is useful to confirm PCR and culture results and to correlate pro- and anti-inflammatory host immune responses with lesion severity and cytokine gene or protein expression. Histologic assessment with a silver stain has a role in identifying gastric bacteria with spiral morphology consistent with *H. pylori* but is a relatively insensitive technique and lacks specificity. A variety of spiral bacteria colonizing the lower bowel of mice can be observed in the stomach, particularly if gastric atrophy develops, and these species are not morphologically distinct at the level of light microscopy either in the stomach or lower bowel. Other less commonly used techniques to localize *H. pylori* in tissues include immunohistochemistry using labeled polyclonal antisera or in situ hybridization for *H. pylori* rRNA. In this chapter, we will summarize strategies to allow initiation of experiments with helicobacter-free mice and then focus on PCR and ELISA techniques to verify and quantify *H. pylori* infection of research mice.

Key words: *Helicobacter pylori*, Mice, ELISA, Quantitative PCR

1. Introduction

Gastric helicobacter infections modeled in mice include mouse-adapted *Helicobacter pylori* SS1 (1), *Helicobacter felis* for its ability to induce robust gastritis (2) and *Helicobacter heilmannii* for its

JeanMarie Houghton (ed.), *Helicobacter Species: Methods and Protocols*, Methods in Molecular Biology, vol. 921, DOI 10.1007/978-1-62703-005-2_19, © Springer Science+Business Media, LLC 2012

induction of MALT lymphoma (3). To most closely model *H. pylori* gastritis and the sequelae of gastric atrophy and cancer in humans, *H. pylori* SS1 is commonly used due to the restricted scope of *H. pylori* strains that will colonize mice. Included here are PCR and serology methods within the capability of most laboratories for verifying the helicobacter-free infection status of research mice pre-dosing and verifying and quantifying *H. pylori* SS1 infection post-dosing. PCR and serology techniques are similar for EHS as well as gastric helicobacters with the exception that fecal PCR has not been a reliable technique in our laboratory for detecting *H. pylori* infection of mice. The value of histologic techniques, such as silver stain, immunohistochemistry and in situ hybridization, and culture as screening techniques is summarized but details of these latter techniques are not within the scope of this chapter.

1.1. Strategy to Verify Helicobacter Infection Status of Research Mice

Investigators need to be aware that the *Helicobacter* genus currently includes at least nine formally named enterohepatic species (EHS), as well as a number of novel isolates yet to be formally named, that colonize the cecum and colon of mice unless specific precautions are taken to exclude them from research colonies (4, 5). Helicobacter-free mice are routinely available from commercial vendors but endemic EHS infection of mice maintained in academic facilities is common (4). Investigators using *H. pylori* mouse models should confirm that their mice are free of EHS as impact of concurrent helicobacter-associated disease has been demonstrated in at least two mouse models (6, 7). Commercial vendors maintain health profiles of their mouse colonies on their Web sites. Note that specific pathogen free (SPF) health status for *Helicobacter* spp. should be verified using *Helicobacter* genus PCR primers and not be limited to *Helicobacter* species-specific exclusion. Mice naturally infected with EHS should be embryo transfer rederived, confirmed helicobacter-free by genus-inclusive PCR, and subsequently maintained in a helicobacter-free barrier facility.

To verify that research mice are helicobacter-free, feces (for example, from live mice, prior to *H. pylori* experimental infection) or a mucosal scraping of the cecal–colic junction (for example to confirm the status of EHS infection at the completion of an experiment) collected aseptically at necropsy should be PCR amplified using *Helicobacter* genus-specific primers (8). If helicobacter speciation is important, PCR products can be sequenced or nested PCR performed. Using *Helicobacter* species-specific primers to amplify the PCR product from the initial genus-level PCR will confirm the presence or absence of specific EHS as well as enhance the overall sensitivity of the PCR screen for EHS (9). For additional evidence and further characterization of EHS infection, feces or cecal scrapings should be stored at –70°C in Brucella broth containing 30% glycerol pending microaerobic culture by an appropriately experienced microbiologist (see Chapter on culture techniques).

Warthin-Starry or Steiner silver stains can be used to localize EHS in the liver parenchyma and biliary system of susceptible mouse strains or gastric helicobacters (*H. pylori*, *H. felis*) in the stomach crypts of experimentally challenged mice. Silver staining of lower bowel to screen for EHS is unrewarding because other morphologically similar bacteria may also be visualized with the silver stain. In the liver of EHS-infected mice, helicobacters such as *H. hepaticus* or *H. bilis* are localized between hepatocytes within biliary cuniculi and can be difficult to visualize if the inflammation is significant. Immunohistochemistry using polyclonal antibodies to detect EHS in mouse livers (10) and *H. pylori* in mouse stomachs have been described (11) as has in situ hybridization for whole cell detection of *H. pylori* in mouse stomach (11) using probes developed for detecting rRNA of *H. pylori* in human gastric biopsies (12).

Commercial serology for helicobacter infection of rodents is not available and lack of seroconversion is not definitive evidence of a helicobacter-free health status, particularly in transgenic mice with potential immunodeficiency. In contrast, successful experimental infection of immunocompetent mice is readily supported by "in-house" positive ELISA serology to helicobacter antigens and has been shown to be an assay with sensitivity exceeding 90%, although with variable specificity, particularly if mice are infected concurrently with multiple EHS (13). The ELISA methods described here are intended to confirm seroconversion of experimentally dosed mice and to characterize the IgG subclass response to chronic *H. pylori* (or *H. felis*) colonization, as B cell isotype class switch recombination reflects polarization of the pro- and anti-inflammatory T cell subset responses. IgG1 and IgG2a/2c (14) subclass responses will typically support pathology findings and cytokine profiles in helicobacter studies (6, 7). As a research tool, serology is inexpensive, noninvasive and allows for serial sampling of individual animals.

2. Materials

2.1. Helicobacter Genus-Specific 16S rDNA-Based PCR Assays for Detecting Helicobacters in Feces and Tissues

1. High Pure PCR Template Preparation Kit (Roche Applied Science Part No.: 11 796 828 001).
2. QIAamp DNA Stool Mini Kit (Qiagen Inc., Valencia, CA Part No.: 51504).
3. RNase-Free disposable pellet pestles with microtubes.
4. Fresh feces or murine tissues which have been stored in sterile 2 mL microtubes at –20°C prior to DNA extraction.

5. Lysozyme at 10 mg/mL in 10 mM Tris–HCl (pH 8.0).
6. Sterile PBS (adjust pH to 7.4 with HCl): 137 mM NaCl, 2.7 mM KCl, 10 mM Na_2HPO_4, 2 mM KH_2PO_4.
7. PuReTaq™ Ready-To-Go™ PCR beads (GE Healthcare).
8. Expand High Fidelity PCR System (Roche Applied Sciences, Indianapolis, IN): Expand High Fidelity Enzyme mix with 10× Reaction Buffer.
9. Agarose.
10. 1× TAE buffer: 40 mM Tris-acetate, 1 mM EDTA (pH 8.0).
11. 1× TE buffer: 10 mM Tris–HCl (pH 8.0), 1 mM EDTA (pH 8.0).
12. Ethidium bromide.
13. Prepare a solution containing both forward and reverse primers (5 μM of each primer) as 10× stock (see Note 1).
14. Ure-B primers (Applied Biosystems). Original stocks of primers and probe at 100 μM; 20× stock: a solution containing 4 μM of both primers and the probe (see Note 2).
15. TaqMan 7500 FAST detection system.
16. MicroAmp® Fast Optical 96-Well Reaction Plate.
17. TaqMan® Fast Universal PCR Master Mix (2×).
18. Adhesive film.
19. FAM-labeled 18S rRNA gene-based primer/probe solution.
20. Electrophoresis loading buffer (6×): 0.25% (w/v) bromophenol blue, 0.25% (w/v) xylene cyanol FF, 30% (v/v) glycerol in ddH_2O.

2.2. ELISA Materials

1. Bacterial sonicator (Misonix Ultrasonic Processor Model S-4000, QSonic LLC, Newtown, CT, or equivalent).
2. Phase microscope to confirm adequate cell lysis (see Notes 3 and 4).
3. *N*-octyl-beta-glucopyranoside.
4. Ultracentrifuge capable of generating 100,000 × *g* for 1 h at 4°C (Sorvall Ultra Pro 80, Thermo Scientific) or equivalent.
5. Compatible ultracentrifuge tubes.
6. Mineral oil for balancing centrifuge tubes.
7. 4 in. section of dialysis tubing with a molecular weight cutoff of 12–14,000 kDa.
8. Two locking, compression clamps designed to seal dialysis tubing.
9. Spectrophotometer.
10. Lowry kit.
11. Sorvall RC-5B Refrigerated Superspeed Centrifuge (or equivalent).

12. Bovine serum albumen (BSA).
13. ELISA plate blocking buffer (volumes of 1 and 2% BSA in PBS).
14. Timer.
15. 37°C incubator.
16. Antigen coating buffer: 1.59 g Na_2CO_3, 2.93 g $NaHCO_3$, in 1 L of ddH_2O, pH of 9.6 (see Note 5).
17. Immulon II High Binding (protein) (designated as 2HB) 96-well plates (Thermo Scientific).
18. Biotinylated secondary (detection) antibodies including polyclonal goat anti-mouse IgG and monoclonal rat anti-mouse antibodies produced by clones A85-1, R19-15 and 5.7 (BD PharMingen) for detecting IgG1, IgG2a, and $IgG2a^b$ (also known as IgG2c), respectively (see Note 6).
19. Extravidin peroxidase (Sigma) diluted 1:2,000 in 1% BSA in PBS.
20. 2,2′-azinobis (3-ethylbenzthiazoline-6-sulfonic acid) diammonium salt (ABTS®) substrate.
21. Automated ELISA plate washer with six cycle wash of 300 μL per well.
22. PBS-Tween (0.05%).
23. ELISA plate reader such as Dynatech MR7000 (Dynatech Laboratories, Inc., Chantilly, VA) capable of reading wavelength matched to color substrate (such as 405 nm for ABTS®).
24. Plastic wrap.
25. 1% sodium dodecyl sulfate (SDS).

3. Methods

3.1. PCR for Helicobacter Infection in Animals

Unless otherwise specified, all procedures are conducted at room temperature and all reagents and solutions are prepared with deionized ddH_2O.

3.1.1. Preparation of PCR DNA Templates (see Note 7)

1. DNA from murine tissues should be isolated using the High Pure PCR Template Preparation Kit following protocol 2.4 in the kit manual. Bacterial DNA from cultivated helicobacter cells should be isolated using the High Pure PCR Template Preparation kit following protocol 2.6 included in the kit directions (see Note 8).
2. Isolate bacterial DNA from cultivated helicobacter as a control (see Note 9).
3. Prepare DNA from feces for screening mice for EHS (see Note 10) according to the protocol described in the QIAamp DNA Stool Mini Kit (see Note 11).

3.1.2. PCR Detection of Helicobacter DNA Using the Genus-Specific Primers (see Note 12) Using Ready-to-Go PCR Beads (see Note 13)

1. Add ~50–500 ng of DNA template, 2.5 μL of 10× primer stock for C97/C05 or C98F/H3A-20 and ddH_2O up to a final volume of 25 μL in a 0.5-mL microtube containing PCR beads. Mix gently with pipetting and centrifuge the reaction mixture for 10 s at 9,000 × *g*.
2. Place the samples into a PCR machine fitted with a heated lid and run the assay using an initial incubation step of 3 min at 94°C followed by cycle program of 1 min at 94°C, 1 min at 58°C, and 2 min at 72°C for 35 cycles followed by 8 min at 72°C and holding at 6°C pending removal of the samples.
3. Mix 10 μL of each PCR sample with 2 μL of 6× electrophoresis loading buffer and load samples on a 1.2% agarose gel. PCR products are separated by electrophoresis at 100 V for ~60 min in 1× TAE buffer followed by staining in ddH_2O containing 0.5 μg/mL ethidium bromide for 15–30 min. After destaining with ddH_2O for at least 15 min, DNA bands can be visualized under UV light.

3.1.3. PCR with High Fidelity PCR System (see Note 14)

1. Set up the reaction mixture in a total volume of 50 μL: template DNA (generally 5 μL containing 50–500 ng template DNA), 1 μL of dNTP mixture (10 mM each of dATP, dCTP, dGTP, dTTP), 5 μL of 10× primer stock for C97/C05 or C98F/H3A-20, 5 μL of Expand High Fidelity Buffer 2, 0.75 μL Expand High Fidelity Enzyme mix (2.6 U/reaction), with balance to 50 μL with ddH_2O. The thermocycling program is the same as described step 2 of Subheading 3.1.2. *Using Ready-to-Go PCR beads.*

3.1.4. Quantitative PCR (qPCR) for Estimating H. pylori Strain SS1 Colonization Levels in Tissues of Experimentally Infected Mice (see Note 15)

1. Add 5 μL of each template DNA into duplicate wells of a MicroAmp® Fast Optical 96-Well Reaction Plate (see Note 16). In two non-template control (NTC) wells, add 5 μL ddH_2O in place of DNA for ruling out DNA contamination.
2. Create a standard curve using serial tenfold dilutions of *H. pylori* SS1 genomic DNA (see Note 17). 5 μL of each standard (equal to 10^6, 10^5, 10^4, 10^3, 10^2, 10^1, respectively, of SS1 genome copies) is added in duplicate to the MicroAmp® Fast Optical 96-Well Reaction Plate.
3. Carefully transfer 15 μL of a reaction mixture containing 10 μL of 2× Master Mix, 1 μL of 20× Primers/Probe mix, and 4 μL ddH_2O (see Note 18) into each well containing controls or samples (see Note 19).
4. Seal the plate with the kit-supplied adhesive film and compression pad. Centrifuge plates briefly at 1,050×*g* in a Labofuge 400 (Heraeus Instruments) before loading it into the TaqMan 7500 FAST detection system.
5. Warm the TaqMan 7500 for at least 15 min then follow the manufacturer's software instructions for setting up an assay

plate, including defining baseline and data analyses. Run the assay in the 7,500 Fast using the default setting (requires about ~40 min).

6. To normalize genomic copies of *H. pylori* SS1 between samples, quantities of mouse DNA in each sample are first measured using the 18S rRNA gene-based primer/probe mixture supplied by Applied Biosystems as described in steps 1–5. A standard curve is generated from serial dilution of DNA isolated from the cecum (tissue) of *Helicobacter*-free mice as follows in pg: 10^5, 10^4, 10^3, 10^2, 10^1. Genomic copies of *H. pylori* SS1 may then be reported as copies of *H. pylori* per μg mouse DNA.

3.2. Antigen Preparation

3.2.1. Sonication of Helicobacter Cells for Antigen

1. Harvest helicobacter cells from ten confluent blood agar culture plates or from a 250 mL broth suspension. Pellet the broth suspension by centrifugation using Sorvall RC-5B Refrigerated Superspeed Centrifuge (or equivalent) set at 6,000 rpm (to achieve approximately 5,000 × *g*) for 10 min at 4°C. Resuspend harvested cells in 10 mL sterile PBS (see Note 20). Repeat PBS wash two more times with final resuspension in approximately 1 mL sterile PBS (should be turbid). Confirm culture purity by gram stain and phase microscopy.
2. Sonicator instrument (use in a biocontainment hood) settings for duration, amplitude, and number of cycles need to be empirically established. The goal is to lyse the majority of cells without overheating the suspension. Keep the suspension in an ice bath during the entire sonication cycle to prevent overheating of proteins. Using the Misonix sonicator, an example of a sonication cycle is 15–30 s of sonication using the Microtip™ set at an amplitude of 40%, chilling the suspension for an additional 20–60 s on ice and then repeating the entire cycle 4–5 more times. Turbid bacterial suspensions will increase in clarity as cells are effectively lysed which should be confirmed by phase microscopy. Sterile technique is not essential but sterile PBS and tubes should be used and the sonicator probe should be cleaned with 70% ethanol prior to use. Measure total protein concentration (Lowry kit) and store sonicate at −20°C and aliquot into 1 mL maximum volumes (to avoid freeze–thaw damage). Typical protein concentrations achieved are up to 1,500 μg/mL.

3.2.2. Preparation of Outer Membrane Proteins for Antigen

This technique was adapted from a previous publication (15).

1. Perform step 1 as under Sonication method except resuspend the final pellet (following the last PBS wash) in 4 mL of 1% *N*-octyl-beta-glucopyranoside (Sigma) (40 mg in 4 mL of PBS).
2. Incubate suspension at room temperature for 30 min. Vortex every 10 min to encourage membrane digestion.

3. Pipette suspension into ultracentrifuge tubes and bring tube volume close to the top of the tube with mineral oil (to prevent tube collapse during ultracentrifugation) and balance tube total weights with mineral oil.
4. Ultracentrifuge at 100,000 × *g* for 1 h at 4°C.
5. In the interim, soak a 4 in. section of dialysis tubing in PBS at room temperature to wet membrane.
6. After ultracentrifugation, pipette off the majority of the mineral oil and discard. Insert pipette tip down through any remaining mineral oil and remove the middle, clear to amber, fluid layer which contains the solubilized OMPs. The pelleted material consists of cytosolic debris and should be discarded.
7. Approximately 1 in. from the end of a 4 in. section of pre-wetted dialysis tubing, firmly apply a locking, compression clamp. Gently pipette (to avoid membrane tear) the outer membrane proteins (OMP) solution into the dialysis tubing followed by firmly applying a second clamp just above the fluid layer. Leave 1 in. of dialysis tubing protruding from both clamped ends to ensure a tight seal.
8. Place dialysis tubing containing OMP solution into 2 L of PBS for 12 h at 4°C. Decant and replace the PBS and dialyze for an additional 6 h. Times and temperatures are empirical and using a stir bar is optional. Slight modifications are not known to alter ELISA performance.
9. Measure total protein concentration (Lowry kit) and aliquot into 1 mL maximum volumes for use or storage at −20°C (to avoid freeze–thaw damage). Typical protein concentrations achieved are lower than by the sonication method but optimally will be 400–1,000 μg/mL.

3.3. Mouse Serum ELISA to Detect Helicobacter-Specific IgG or Isotypes

The following method describes an endpoint ELISA performed at a serum dilution of 1:100. It can be adapted to measuring titers by serial twofold dilution of sera.

One Day Prior to Performing the Assay

1. Design a plate map (see Note 21).
2. Prepare 2% BSA in PBS blocking solution (see Note 22) and carbonate buffer.
3. Coat 96-well Immulon 2 HB plate(s) with 1 μg/mL (for IgG) or 10 μg/mL (for IgG isotypes) of sonicate or OMP antigens in carbonate buffer. Requires 10 mL of either protein concentration per 96-well plate mixed in carbonate. Pipette 100 μL into every well according to the plate map. Cover plate(s) with plastic film to prevent dehydration and incubate at 4°C overnight.

The Following Morning (the Day that ELISA is Performed)

4. Wash plates 6× with 300 μL per well of PBS-0.05% Tween washing buffer. After last wash cycle, invert plate(s) with a sharp motion onto a fresh paper towel to absorb residual wash buffer.
5. Pipette 200 μL of 2% BSA in PBS blocking buffer into every well, Cover plate(s) with plastic film and incubate at 37°C for 1 h.
6. During blocking step, dilute sera 1:100 in 1% BSA in PBS. Make sufficient volume of diluted sera (100 μL of diluted sera per well) to accommodate the plate map with 100–200 μL extra volume of each sample to adjust for pipetting error. For example, add 5 μL serum sample into 495 μL 1% BSA in PBS so each sample can be tested in triplicate (requires 300 μL of diluted sera), leaving 200 μL for pipetting error.
7. Wash plates 6× with 300 μL per well of PBS-0.05% Tween washing buffer. After last wash cycle, invert plate(s) with a sharp motion onto a fresh paper towel to absorb residual wash buffer.
8. Apply 100 μL of samples (or 1% BSA in PBS to control blank wells) following the plate map. Cover plate(s) with plastic film and incubate at 37°C for 1 h.
9. Wash plates 6× with 300 μL per well of PBS-0.05% Tween washing buffer. After last wash cycle, invert plate(s) with a sharp motion onto a fresh paper towel to absorb residual wash buffer.
10. Following plate map, apply 100 μL of biotinylated detection antibody diluted 1:2,000 in 1% BSA in PBS (goat anti-mouse IgG, monoclonal rat anti-mouse IgG1, 2a, or 2c). Note that secondary (detection) antibodies are also pipetted into blank wells according to plate map. Cover plate(s) with plastic film and incubate at 37°C for 1 h.
11. Wash plates 6× with 300 μL per well of PBS-0.05% Tween washing buffer. After last wash cycle, invert plate(s) with a sharp motion onto a fresh paper towel to absorb residual wash buffer.
12. Apply 100 μL of Extravidin-peroxidase diluted 1: 2,000 to every well designated as blank or sample. Cover plate(s) with plastic film and incubate at 37°C for 30 min.
13. During this incubation (step 9), mix sufficient ABTS solutions labeled A and B (1:1 ratio) to yield 100 μL per well (10 mL per plate). Mix in a glass bottle, store ABTS in the dark and let reach room temperature before use.
14. Wash plates 6× with 300 μL per well of PBS-0.05% Tween washing buffer. After last wash cycle, invert plate(s) with a sharp motion onto a fresh paper towel to absorb residual wash buffer.
15. Apply 100 μL ABTS to every well designated as blank or sample and incubate uncovered (no plastic film) at room temperature for 30 min in the dark.

16. Optionally, add 100 μL 1% SDS as a stop solution; does not change wavelength reading. This stop solution is specific to ABTS and other stop solutions do change the required reading wavelength for other color substrates.
17. Using an ELISA reader, measure optical densities at 405 nm (specific for ABTS substrate) at 30 min. Some ELISA readers have dual wavelength capability and will simultaneously read a distant wavelength such as 590 nm to correct for background optical distortions. Adjust time based on color development, record OD at different time points if indicated.

4. Notes

1. *Helicobacter* genus-specific PCR primers C97 (Forward: 5′-GCTATGACGGGTATCC-3′ corresponding to positions 276–291) and C05 (Reverse: 5′-ACTTCACCCCAGTCG CTG-3′ corresponding to positions 1,478–1,495) were derived from the 16S rRNA gene for amplifying a ~1,200 bp *Helicobacter*-specific PCR product as previously described (8). A pair of internal primers, C98F (Forward: 5′-TGGTGTAGGGGTAA AATC-3′ corresponding to positions 681–698) and H3A-20 (Reverse: 5′-GCCGTGCAGCACCTGTTTTC-3′ corresponding to positions 1,007–1,026), are used for nested PCR based on the C97/C05-amplified products (see Note 12) (16). Positions of the primer sequences are in reference to the *H. pylori* 16S rRNA sequence. Primers are synthesized and SDS-PAGE-purified by Integrated DNA Technologies, Inc. Store the primers at 100 μM in sterile deionized ddH_2O as stock at −70°C.
2. *ureB*-based primers and a probe were developed and used for quantification of *H. pylori* SS1 as previously described (17). Forward primer, 5′-CAAAATCGCTGGCATTGGT-3′; Reverse primer, 5′-CTTCACCGGCTAAGGCTTCA-3′; probe 5′-AACAA AGACATGCAAGATGGCGTTAAAAACA-3′, which is labeled with six FAM dye at 5′-end and with MGB (quencher) at 3′-end.
3. Preparation of antigens from *H. pylori* cultures should be performed in a biohazard containment hood (except for centrifugation steps). Reagents inclusive of secondary (detection) antibodies can be prepared and used at room temperature for the assay duration but otherwise should be stored at 4°C to avoid potential freeze–thaw damage. Sample sera should be stored at −20°C and then thawed and diluted in 1% BSA in PBS just prior to the assay. Consider avoiding freeze–thaw damage to sample sera by aliquoting samples when first collected.

4. Sonication produces a whole cell lysate whereas the OMP method theoretically selects for outer cell wall antigens and discards the more highly conserved, cytosolic proteins. In theory, OMP should yield an assay with higher specificity but this does not have practical significance when seroconversion is being evaluated in mice experimentally infected with a gastric helicobacter and are otherwise SPF for EHS. The protein yield from the OMP method is lower due to the nature of detergent digestion in a 4 mL volume and sample loss inherent in centrifugation steps.
5. This has a 2-week shelf life.
6. Based on mouse genetics, specific detection antibodies must be used to accurately quantify Th1-associated subclass responses because C57BL/6, C57BL/10, and NOD mice do not express the IgG2a isotype and instead produce the $IgG2a^{b}$ isotype, also known as IgG2c (14).
7. Preparation of DNA from feces or tissues used to screen mice for EHS or from *H. pylori*-infected tissues should be performed in a biohazard containment hood (except for centrifugation steps). Some EHS of mice may have zoonotic potential (18, 19).
8. For increasing DNA yield from tissue, homogenize each sample with a pestle in a 1.5 mL microtube prior to incubation with proteinase K as described in protocol 2.4 included in the kit directions.
9. Our laboratory uses High Pure PCR Template Preparation kit following protocol 2.6.
10. Minor modifications to the kit directions include: In step 1 of kit manual, we use 2–3 fecal pellets instead of the suggested 180–220 mg of feces; the fecal pellets are homogenized in 1.4 mL of kit-supplied buffer ASL. In step 6 of kit manual, one-half of an inhibitEX tablet is used rather than the suggested whole tablet. Note that fecal PCR to detect *H. pylori* has not been a reliable diagnostic technique in our laboratory.
11. (Pages 15–18 for isolation of pathogen DNA from stool) with minor modifications (see Note 10).
12. Measure quality of DNA samples based on a ratio of spectrophotometry wavelengths (usually A260/A280 ≥1.6) using a standard spectrophotometer or the more recently available Nanodrop technology (Thermo Scientific) that measures DNA quality using very small volumes (0.5 μL). Adjust tissue or fecal DNA concentrations to approximately 25–50 ng/μL with TE buffer. Thaw and mix DNA samples before loading them onto the plate. For increasing helicobacter detection sensitivity, nested PCR can be performed following the protocols described in Subheadings 3.1.2 or 3.1.3 by using 2 μL of the C97/C05-amplified mixture (obtained using Subheadings 3.1.2 or 3.1.3) as nested PCR template and 10× primer stock of C98F/H3A-20.

13. PCR beads contain all required reagents for PCR except for primers and template DNA and are supplied in individual PCR tubes. Advantages of using PCR beads include minimizing variability in reagent preparation and minimizing DNA contamination. PCR beads are generally used for production of amplicon products <2 kb in size. In some samples, nonspecific amplicons may be generated along with the amplicon of interest.
14. High Fidelity PCR System yields PCR products up to 5 kb in size with high specificity and sequence fidelity. PCR products used for subsequent cloning and sequencing should be produced using this system to minimize introduction of error nucleotides during PCR. Disadvantages of the High Fidelity PCR System include increased expense, time for preparing PCR reactions and increased risk of DNA contamination.
15. Due to extensive nucleotide sequence diversity within allelic genes among *H. pylori* strains, these primers and probe may not adequately hybridize to non-SS1 *H. pylori* strains.
16. The manufacturer recommends use of quadruple replicates per sample. We found that use of duplicate samples reduced cost without compromising results.
17. Based on the average size (~1.66 mb) of two sequenced *H. pylori* strains 26695 (3) and J99 (4), two fentograms of *H. pylori* SS1 DNA is approximately equal to one genome copy.
18. When pipetting samples into wells, avoid creating air bubbles that will interfere with fluorescence readings. Carefully pipette the reaction mixture along the wall of each well and allow the mixture to settle to the bottom by gravity.
19. The reaction mixture of primers/probe for all samples can be prepared in a single tube and distributed into individual wells using an eight-channel pipette. Prepare extra reaction mixture volume to compensate for pipetting error, e.g., one extra reaction volume per eight wells.
20. To prepare antigen from broth culture, inoculate 250 mL of Brucella broth containing 5% fetal calf serum with helicobacter cells harvested from one confluent blood agar plate. Incubate overnight at 37°C to achieve a turbidity reading measured by spectrophotometry of 1.0 at 600 nm wavelength. 250 mL will generate a cell pellet equivalent to approximately 2×10^{11} cells.
21. Design a plate map to maximize use of multichannel pipettes to efficiently apply reagents. Assign plate wells for triplicate control (i.e., blank or background wells without sera) and triplicate samples. For a total IgG assay, designate well 1–3 in row A as blanks and designate the remaining 93 wells for samples to be pipetted in triplicate, oriented either in a horizontal or vertical direction to fill the plate (can thus accommodate 31 test sera). For IgG isotype measurements, each isotype requires

its own triplicate blank wells reserved in row A. An efficient design is to designate row A wells 1–3 for IgG blanks, row A wells 4–6 for IgG1 blanks, and row A wells 7–9 for IgG2a or IgG2c blanks (see Note 6). Each sample can then be pipetted horizontally across nine wells using a multichannel pipette without having to change pipette tips. Remaining samples are then similarly pipetted into rows B through H under their respective blanks. In this design, columns 10–12 would be left empty. The plate map should indicate the antigen coating concentration (1 μg/mL for IgG, 10 μg/mL for isotypes) and where each individual detection antibody (anti-IgG, anti-IgG1, or anti-IgG2a/2c) will be pipetted. After application of detection antibodies, every well receives the same reagents (i.e., Extravidin peroxidase, ABTS).

22. Because BSA is hydrophobic, make 2% BSA in PBS blocking buffer on the day prior to the ELISA by adding 2 g of BSA to every 100 mL of PBS required and store at 4°C overnight. Mixing is not necessary and unnecessarily creates protein foam. The blocking step will require 20 mL of blocking buffer per plate. Make excess blocking solution to conveniently create 1% BSA in PBS during the blocking incubation step the following day to accommodate dilution of sera and other reagents (simply adding equal volume of PBS to the 2% BSA in PBS left over from the blocking step to create 1% BSA in PBS).

References

1. Lee A, Orourke J, Deungria MC, Robertson B, Daskalopoulos G, Dixon MF (1997) A standardized mouse model of *Helicobacter pylori* infection: introducing the Sydney strain. Gastroenterology 112:1386–1397
2. Lee A, Fox JG, Otto G, Murphy J (1990) A small animal model of human *Helicobacter pylori* active chronic gastritis. Gastroenterology 99:1315–1323
3. O'Rourke JL, Dixon MF, Jack A, Enno A, Lee A (2004) Gastric B-cell mucosa-associated lymphoid tissue (MALT) lymphoma in an animal model of '*Helicobacter heilmannii*' infection. J Pathol 203:896–903
4. Taylor NS, Xu S, Nambiar P, Dewhirst FE, Fox JG (2007) Enterohepatic *Helicobacter* species are prevalent in mice from commercial and academic institutions in Asia, Europe, and North America. J Clin Microbiol 45:2166–2172
5. Whary MT, Fox JG (2004) Natural and experimental *Helicobacter* infections. Comp Med 54:128–158
6. Lemke LB, Ge Z, Whary MT, Feng Y, Rogers AB, Muthupalani S, Fox JG (2009) Concurrent *Helicobacter bilis* infection in C57BL/6 mice attenuates proinflammatory *H. pylori*-induced gastric pathology. Infect Immun 77:2147–2158
7. Stoicov C, Whary M, Rogers AB, Lee FS, Klucevsek K, Li H, Cai X, Saffari R, Ge Z, Khan IA, Combe C, Luster A, Fox JG, Houghton J (2004) Coinfection modulates inflammatory responses and clinical outcome of *Helicobacter felis* and *Toxoplasma gondii* infections. J Immunol 173:3329–3336
8. Fox JG, Dewhirst FE, Shen Z, Taylor NS, Paster BJ, Ericson RL, Lau CN, Correa P, Araya JC, Roa I (1998) *Hepatic Helicobacter* species identified in bile and gallbladder tissue from Chileans with chronic cholecystitis. Gastroenterology 114:755–763
9. Maurer KJ, Ihrig MM, Rogers AB, Ng V, Bouchard G, Leonard MR, Carey MC, Fox JG (2005) Identification of cholelithogenic enterohepatic helicobacter species and their role in murine cholesterol gallstone formation. Gastroenterology 128:1023–1033

10. Li X, Fox JG, Whary MT, Yan L, Shames B, Zhao Z (1998) SCID/NCr mice naturally infected with *Helicobacter hepaticus* develop progressive hepatitis, proliferative typhlitis, and colitis. Infect Immun 66:5477–5484
11. Fox JG, Wang TC, Rogers AB, Poutahidis T, Ge Z, Taylor N, Dangler CA, Israel DA, Krishna U, Gaus K, Peek RM Jr (2003) Host and microbial constituents influence *Helicobacter pylori*-induced cancer in a murine model of hypergastrinemia. Gastroenterology 124:1879–1890
12. Trebesius K, Panthel K, Strobel S, Vogt K, Faller G, Kirchner T, Kist M, Heesemann J, Haas R (2000) Rapid and specific detection of *Helicobacter pylori* macrolide resistance in gastric tissue by fluorescent in situ hybridisation. Gut 46:608–614
13. Whary MT, Cline JH, King AE, Hewes KM, Chojnacky D, Salvarrey A, Fox JG (2000) Monitoring sentinel mice for *Helicobacter hepaticus*, *H. rodentium*, and *H. bilis* infection by use of polymerase chain reaction analysis and serologic testing. Comp Med 50:436–443
14. Martin RM, Brady JL, Lew AM (1998) The need for IgG2c specific antiserum when isotyping antibodies from C57BL/6 and NOD mice. J Immunol Methods 212:187–192
15. Pronovost AD, Rose SL, Pawlak JW, Robin H, Schneider R (1994) Evaluation of a new immunodiagnostic assay for *H. pylori* antibody detection: correlation with histopathological and microbiological results. J Clin Microbiol 32:46–50
16. Bohr UR, Primus A, Zagoura A, Glasbrenner B, Wex T, Malfertheiner P (2002) A group-specific PCR assay for the detection of *Helicobacteraceae* in human gut. Helicobacter 7:378–383
17. Maurer KJ, Rogers AB, Ge Z, Wiese AJ, Carey MC, Fox JG (2006) *Helicobacter pylori* and cholesterol gallstone formation in C57L/J mice: a prospective study. Am J Physiol Gastrointest Liver Physiol 290: G175–G182
18. Fox JG (1997) The expanding genus of *Helicobacter*: pathogenic and zoonotic potential. Semin Gastrointest Dis 8:124–141
19. Boutin SR, Shen Z, Roesch PL, Stiefel SM, Sanderson AE, Multari HM, Pridhoko EA, Smith JC, Taylor NS, Lohmiller JJ, Dewhirst FE, Klein HJ, Fox JG (2010) *Helicobacter pullorum* outbreak in C57BL/6NTac and C3H/HeNTac barrier-maintained mice. J Clin Microbiol 48:1908–1910

Chapter 20

Mouse Models of *Helicobacter*-Induced Gastric Cancer: Use of Cocarcinogens

Richard L. Ferrero, John E. Wilson, and Philip Sutton

Abstract

The human pathogen *Helicobacter pylori* causes inflammation in the stomach of infected hosts, leading in some cases to the development of gastric cancer. Several mouse models have been developed to study *Helicobacter*-induced carcinogenesis with similarities to gastric adenocarcinoma and mucosa-associated lymphoid tissue lymphoma (MALToma) in humans. These models require chronic infection of animals with mouse-colonizing isolates of *H. pylori* or with related gastric *Helicobacter* spp., such as the canine/feline species *Helicobacter felis*. Furthermore, consistent with the known influence of host and environmental factors in human gastric cancer, it is possible to manipulate the type and severity of gastric lesions in mouse *Helicobacter* infection models through the use of different mouse genetic backgrounds and/or by the administration of known cocarcinogens, such as alkylating agents (e.g., *N*-nitroso-*N*-methylurea), or even elevated quantities of dietary salt. Here, we describe protocols for the inoculation of mice with gastric *Helicobacter* spp. and the administration of these cocarcinogens. Furthermore, we will describe the various methodologies used to study gastric inflammation and carcinogenesis in *Helicobacter*-infected animals.

Key words: *Helicobacter pylori*, *Helicobacter felis*, Mouse model, Gastric cancer, Adenocarcinoma, MALT lymphoma, Gastritis, Salt, Cocarcinogens, MNU, MNNG

1. Introduction

Mice have proven to be very useful as models in the study of the pathogenesis, prevention and/or treatment of various types of human disease. This is particularly true in the *Helicobacter* field, where mice experimentally infected with a variety of gastric *Helicobacter* spp. have been shown to reproduce many aspects of human *H. pylori* infection, including the two major forms of gastric cancer, adenocarcinoma and mucosa-associated lymphoid tissue lymphoma (MALToma), which affect gastric gland and lymphoid cell populations, respectively. Indeed, it has been shown that mice infected for long periods with

JeanMarie Houghton (ed.), *Helicobacter Species: Methods and Protocols*, Methods in Molecular Biology, vol. 921, DOI 10.1007/978-1-62703-005-2_20, © Springer Science+Business Media, LLC 2012

either *H. pylori* or the related feline/canine species, *Helicobacter felis* or *Helicobacter heilmannii*, develop the chronic gastritis that is a hallmark of *H. pylori* infection (1, 2). Furthermore, *H. felis* was reported to cause adenocarcinoma in C57BL/6 mice at 24 months post-infection (3), whereas 38% of *H. felis*-infected BALB/c mice were observed to develop MALToma-like lesions at 22 months (4). In C57BL/6 mice, *H. felis* is able to cause the full progression of disease seen in human gastric adenocarcinoma: from chronic gastritis to glandular atrophy, metaplasia, dysplasia, and, ultimately, adenocarcinoma (3). Interestingly, in contrast, mouse-colonizing *H. pylori* isolates induce relatively modest levels of gastritis and though these isolates can induce gland atrophy in certain mouse host backgrounds (e.g., C57BL/6 mice) (2, 5), they do not normally induce neoplastic disease in the murine host. In fact, *H. pylori* infection has only been shown to be associated with preneoplastic or neoplastic disease in mouse strains with a predisposition to gastric cancer (6–8) or, alternatively, in animals that were coadministered cocarcinogens (1, 9–11). Presented below is a description of the methods required for the establishment of gastric *Helicobacter* infection and the administration of cocarcinogens in mice, as well as the various methodologies used to study gastric inflammation and carcinogenesis in these animals.

2. Materials

2.1. Mice

Source, sex, age. Status (gnotobiotic, germ-free, specific pathogen free) Helicobacter-free.

Specific pathogen-free male or female mice, of 6–8 weeks age, can be used for mouse infection studies with either *H. pylori* or *H. felis* (see Note 1).

2.2. Bacterial Strains

The most commonly used *H. pylori* strains in mouse colonization studies are: SS1 ("Sydney strain 1") (12); B128 and its gerbil-adapted variant, B128 7.13 (13); and X47-2AL (14), which was isolated from an *H. pylori*-infected cat (see Note 2). Several other *H. pylori* strains have been described, but these have not gained widespread acceptance (e.g., *H. pylori* HAS-141, SPM326, "Sydney 2000," 245, 256, etc.).

For *H. felis* infection studies in mice, workers have almost exclusively used *H. felis* CS1 ("cat spiral 1"; ATCC49179) (15).

2.3. Preparation of Bacterial Inocula and Cocarcinogen Solutions

1. *Helicobacter* storage medium (500 mL): 125 mL glycerol to 190 mL Brain-Heart Infusion (BHI) broth and 185 mL distilled water. Mix well and verify that pH is between 7.0 and 7.2. Dispense aliquots of medium into glass vials and autoclave at 115°C for 20 min. Alternatively, dispense autoclaved medium into cryotubes (Nunc), as required. Sterile medium can be stored at 4°C.

2. *Helicobacter* solid medium (per 500 mL): dissolve 20 g of Blood Agar Base no.2 (Oxoid), or an equivalent nutrient-rich agar base (see Note 3), by gentle swirling in approximately 500 mL of distilled water, using 1 L volumetric flasks or Schott bottles. Sterilize agar medium by autoclaving at 121°C for 20 min. After cooling the agar in a water bath set to 56°C, add aseptically: 35–50 mL of sterile fresh horse blood (Oxoid), which has been pre-warmed to room temperature and 1 mL of modified "Skirrow's Selective Supplement" (see below and Note 4). Let the plates set on the bench (see Note 5) then store in sealed plastic petri dish bags at 4°C for up to several weeks.
3. Modified Skirrow's Selective Supplement (40 mL): polymyxin B sulfate salt (6.2 mg), vancomycin hydrochloride (250 mg), trimethoprim (125 mg in a small volume of 100% ethanol) and amphotericin B (50 mg) in 40 mL distilled H20. Solutions are prepared in 50 mL sterile polyethylene tubes, which are then wrapped in aluminum foil and stored at 4°C for up to several months.
4. Brain-Heart Infusion (BHI) Broth (per liter). Thoroughly resuspend 37 g BHI broth medium in distilled water by gentle swirling, or by the use of a magnetic stirrer, in a volumetric flask or bottle (see Note 6) and make up to final volume. Distribute the resuspended BHI broth (100 or 400 mL aliquots) in appropriately sized, prerinsed Schott bottles. Sterilize the aliquots of BHI broth by autoclaving at 121°C for 20 min. Cool BHI broth prior to use and store at 4°C.
5. *Helicobacter* liquid culture medium (per 100 mL). *H. pylori* liquid culture medium is prepared in sterile 75 cm^2 tissue culture flasks (Falcon) by adding BHI broth (100 mL) supplemented with Skirrow's Selective Supplement and 10% (v/v) of heat-inactivated fetal calf serum (FCS; Invitrogen) (see Note 7). This medium can be stored at 4°C for several weeks.
6. *H. pylori* solid medium for quantitative bacterial counts. To BA culture medium, add 10 mg/mL nalidixic acid and 200 μg/mL bacitracin (5) (see Note 8).
7. Phosphate-buffered saline (PBS; 0.1 M). Add 4.37 g/L of $NaH_2PO_4{\cdot}2H_2O$, 10.22 g/L Na_2HPO_4 and 8.5 g/L NaCl to distilled water and adjust pH to 7.2–7.4. Make up to final volume, dispense if required, and autoclave at 121°C for 20 min. Store at room temperature.
8. Physiological saline (0.85%). Add 8.5 g of NaCl to distilled water and adjust pH to 7.0. Make up to final volume, dispense if required, and autoclave at 121°C for 20 min. Store at room temperature.
9. Cocarcinogens

(a) *N*-nitroso-*N*-methylurea (MNU). Prepare 240 mg/mL MNU in drinking water (9). Solutions are freshly prepared twice a week in light-shielded bottles.

(b) *N*-methyl-*N′*-nitro-*N*-nitrosoguanidine (MNNG). Prepare 150 mg/mL MNNG in drinking water (1). 750 mg of MNNG is diluted to a concentration of 150 mg/mL in 0.02% dimethylsulphoxide (DMSO) in sterile distilled water. Solutions are prepared in light-shielded bottles and are changed every 2 days (1).

(c) Salt. The levels of dietary salt in mice can be controlled using specially formulated mouse chow containing low or high concentrations of salt: 0.25% and 7.5%, respectively (our laboratory uses Special Formulation Lab Diet, Purina Mills, Richmond, IN) (6, 11).

2.4. Reagents for Tissue Analyses

1. Urease reagent.

   (a) Urea-indole medium (1 L; (5)). Add 3 g l-tryptophane, 1 g KH_2PO_4, 1 g K_2HPO_4, 5 g NaCl, 20 g urea, 10 mL ethanol (95% v/v), and 25 mg phenol red to distilled water. Filter-sterilize solution using a 0.2 μm microfilter and dispense 10 mL aliquots into Macartney bottles or equivalent. Store aliquots at room temperature.

   (b) Hazell's urease reagent (1 L; (15)). Add 0.44 g $NaH_2PO_4{\cdot}2H_2O$, 1.02 g Na_2HPO_4, 20 g urea, 0.5 g phenol red, 0.2 g NaN_3 (sodium azide) to distilled water. Adjust solution to pH 5.7, make up to final volume, and dispense solution into glass bottles. Store at room temperature.

2. Formalin for tissue fixation. Add 4.37 g/L of $NaH_2PO_4{\cdot}2H_2O$ and 10.22 g/L Na_2HPO_4 to approximately 800 mL of distilled water. To this solution, add 100 mL of 40% v/v formaldehyde solution (under hood) and adjust pH to 7.0. Make up to final volume. Store at room temperature.

3. Cryopreservation medium for tissues. For immunohistochemical analyses of tissues, gastric biopsies and other tissues should be placed into molds to which CRYO-OCT Compound is added. These samples are snap-frozen using a liquid nitrogen bath and can be stored at −80°C until processed. Frozen sections are fixed using ice-cold acetone and washed with PBS to remove OCT Compound. Tissues can then be used to perform indirect immunofluorescence and immunohistochemistry.

4. Storage of tissue samples for RNA extraction. Gastric and other tissue biopsy samples for RNA extraction and analysis can be stored in RNA*later*® (Ambion), or equivalent RNA preservation medium.

3. Methods

3.1. Bacteriological Method

3.1.1. Preparation of Stock Cultures of H. pylori and H. felis

1. *H. pylori* and *H. felis* bacteria that have been grown on solid or in liquid medium (see below) are harvested by swabbing or centrifugation, respectively.
2. The bacteria are resuspended in *Helicobacter* storage medium and stored at −80°C, until required.

3.1.2. Growth of H. pylori and H. felis on Solid Media

1. A small aliquot (approximately 100–200 μl) of *H. pylori* or *H. felis* suspension, which has been maintained at −80°C in *Helicobacter* storage medium, is deposited onto a blood agar plate.
2. Streak cultures across the surface of plates with sterile glass Pasteur pipettes or plastic droppers or loops. Do not attempt to generate isolated colonies (see Note 9).
3. Incubate plates with lid uppermost in commercial anaerobe jars, with a petri dish containing distilled water at the base of each jar. Incubate jars at 37°C for 1.5–2 days, for *H. pylori*, and at least 2 days, for *H. felis* (see Note 10).
4. To subculture, streak plates in the same manner as indicated above using sterile loops and add culture plates upright in gas jars, containing a petri dish at the bottom of each jar. Once the bacteria have adapted to in vitro culture by successive passages on agar plates, these can be routinely grown every 1–2 days, always allowing more time for *H. felis*.

3.1.3. Growth of H. pylori and H. felis in Liquid Media

1. Log phase plate cultures of *H. pylori* or *H. felis* (1–1.5 day) should be used to inoculate *Helicobacter* liquid culture medium. Liquid cultures can be grown within either sterile glass Erlenmeyer flasks or, for convenience, within 25 cm^2 vented tissue culture flasks in a final volume of 5–10 mL (or 75 cm^2 vented tissue culture flasks, in a final volume of 10–20 mL).
2. For inoculation of broth media, add thick loopfuls of bacteria from plates directly into the liquid. There should be large clumps of bacteria in the liquid medium.
3. An alternative method is to harvest the bacteria directly from plate cultures by flooding the plates with small volumes of BHI broth (5–10 mL). Use these bacterial suspensions to inoculate fresh *Helicobacter* liquid culture medium.
4. Suspensions should have cell densities corresponding to $OD_{600} = 0.02$ to 0.05, i.e., faintly turbid.
5. Place flasks in Anaerojar (AG0025) anaerobe jars (see Note 10) or equivalent, and incubate at 37°C with shaking at approximately 120–140 rpm.

6. Grow the bacteria to early-to-mid logarithmic phase, i.e., approximately 16–18 h.
7. The viability and general state of the bacteria can be assessed by examination of wet mount preparations under phase contrast microscopy (100× magnification).

3.1.4. Viable Counts on H. pylori Suspensions

1. Perform duplicate or triplicate serial dilutions (10^{-1}—10^{-5}) of *H. pylori* cultures in BHI. For this, prepare 450 μl aliquots of BHI in sterile 1.5 mL volume tubes. To perform the dilutions, transfer 50 μl of the stock bacterial suspension to 450 μl of BHI (10^{-1}).
2. Vortex gently, then transfer with a new tip to another 450 μl of BHI (10^{-2}), etc.
3. Divide pre-dried blood agar plates (see Note 11) into three or four segments. Streak each dilution onto a segment of the divided plates using 10 μl calibrated plastic loops.
4. After having ensured that the inoculated plates are dry, incubate plates in inverted position with petri dish containing water in each gas jar. Incubate cultures for approximately 3–5 days (or until colonies can be seen).
5. Count segment(s) between 10 and approximately 100 isolated colonies. To calculate the number of viable bacteria (in colony-forming units, CFUs) per mL:

   Number of colonies counted × dilution factor × 100 = viable bacteria (CFU) per mL, where the factor of "100" corresponds to the dilution introduced by the use of 10 μl calibrated loops.

3.1.5. Preparation of Bacterial Inocula

1. *H. pylori* or *H. felis* bacteria are taken from −80°C and subcultured twice on solid medium.
2. The bacteria can then be either subcultured a third time on solid agar (1–1.5 days incubation) or grown in liquid broth medium (16–18 h; see Subheading 3.1.2 above). Bacteria should be grown to early-to-mid logarithmic phase (see Note 12). For cultures on solid medium, harvest bacteria directly from the plates using BHI or equivalent broth.
3. The numbers of bacteria in the inocula should initially be approximated by performing total bacterial counts under phase contrast microscopy. For this, count the number of bacteria per field (100× magnification) and determine approximately the number of bacteria/mL using the following guide:

   1 bacterium per field = approximately 10^6 *H. pylori*/mL.

   10 bacteria per field = approximately 10^7 *H. pylori*/mL.

   100 bacteria per field = approximately 10^8 *H. pylori*/mL etc.

4. Following inoculation of the mice, perform viable counts to determine post hoc the actual numbers of bacteria used to inoculate the animals (see Subheading 3.1.3 above).

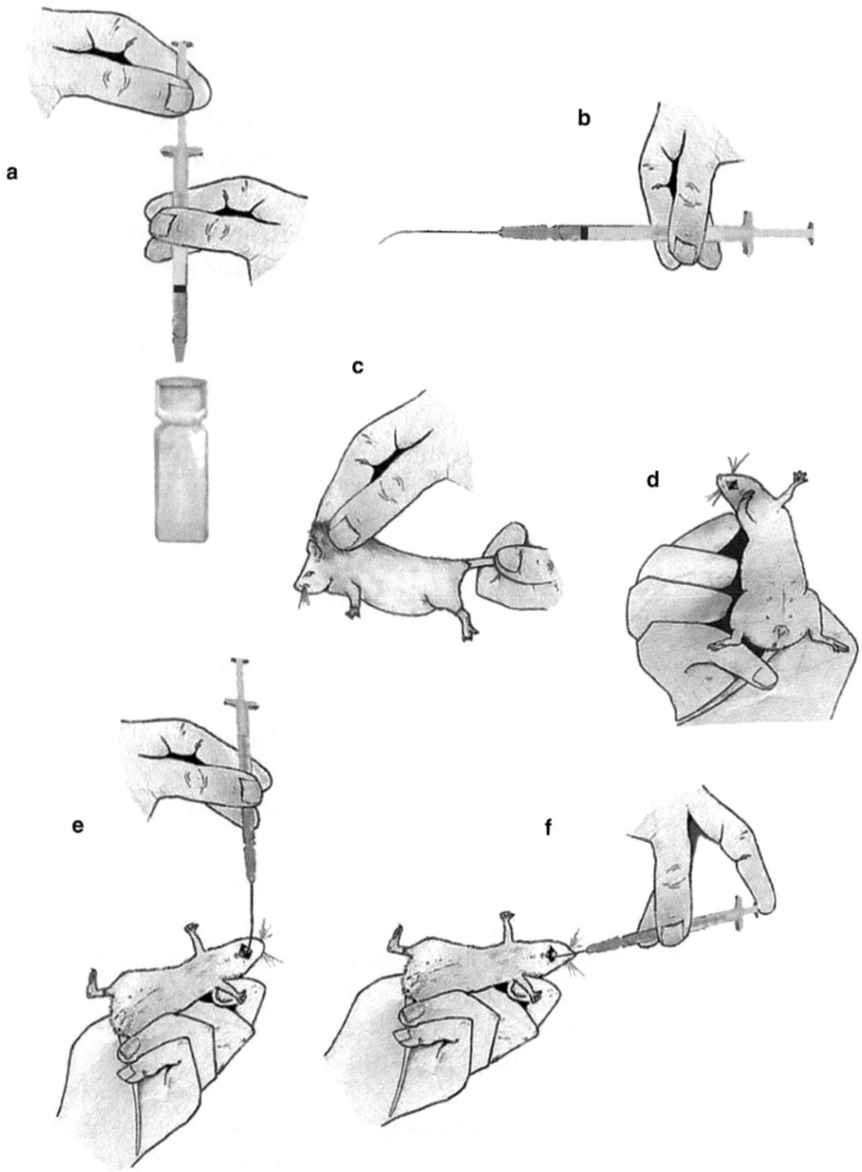

Fig. 1. Intragastric gavage procedure. (**a**) Aspiration of a bacterial suspension into the syringe. (**b**) A flexible catheter is affixed to the syringe containing the bacterial suspension. (**c**) Mouse is restrained by a firm grip at the scruff of the neck and tail. (**d**) Mouse is placed in the supine position. (**e**) The gavage is inserted into the space between the left incisors and molars and guided in a caudal direction towards the esophagus. (**f**) The gavage is inserted down the esophagus into the stomach of the mouse and the suspension administered into the gastric lumen.

3.2. Orogastric Inoculation Procedure

3.2.1. Intragastric Gavage (Fig. 1)

1. This procedure can be performed without any form of anesthesia, or alternatively, with the use of an inhaled anesthetic, such as methoxyflurane or isoflurane. For the latter, small volumes (1–2 mL) of the anesthetic are applied to cotton wool or gauze in screw-top jars, to which are added individual mice.
2. Physically restrain animals by hand and immobilize by application of a firm grip about the scruff of the neck and tail.

3. Insert gavage needle into the space between the left incisors and molars and guided in a caudal direction towards the esophagus.
4. Passage is generally facilitated by the onset of a swallowing reflex as the gavage approaches the pharynx allowing progression into the esophagus.
5. Extend the neck of the mouse to provide a straight line between the esophagus orifice and the cardiac sphincter.
6. Insert gavage needle down the esophagus into the stomach and deliver the specific aliquot (100 μl) (see Note 13).
7. For *H. pylori* SS1 and *H. felis*, single doses of as little as $>2\times10^3$ CFU (5) and $>10^2$ (16), respectively, were shown to be required to infect outbred Swiss mice; however, workers should generally aim to inoculate mice with $\geq10^5$ CFU per mouse (see Note 14).
8. The viability and general state of the bacteria can be assessed by examination of wet mount preparations under phase contrast microscopy.
9. The inocula should not be used if a large proportion of the bacteria (≥10%) are in a nonviable, coccoid form rather than the spiral/helical or bacillary shapes typical of viable bacteria.

3.3. Cocarcinogen Treatment Studies

3.3.1. MNU

1. 240 mg/mL MUN is given via drinking water ad libation five times on alternating weeks over a total period of 10 weeks.
2. 2 weeks later, administer *H. pylori* SS1 by gavage.
3. Euthanize at 38 weeks (i.e., week 50 of the experiment). In this model, 80% of mice given the cocarcinogen and *H. pylori* infection develop gastric adenoma and adenocarcinoma, whereas animals with *H. pylori* infection alone exhibit only chronic atrophic gastritis (9).

3.3.2. MNNG

1. Add MNNG (150 μg/mL) to the drinking water for periods up to 18 months after inoculation with the related gastric *Helicobacter* sp., *H. heilmannii* (1, 9) (see Notes 15 and 16).

3.3.3. Salt

1. 0.25% or 7.5% salt chow is given ad libitum.

3.4. Analyses of Mice Post-Infection/Treatment

3.4.1. Tissue Collection

1. At the conclusion of the experiment, mice are euthanized by either CO_2 inhalation or cervical dislocation, according to the relevant institutional guidelines for animal experimentation.
2. Remove the stomach; open the abdominal cavity by incision with sharp scissors.
3. Sever the pyloric and cardiac sphincters and any attached mesentery (Figs. 2 and 3).

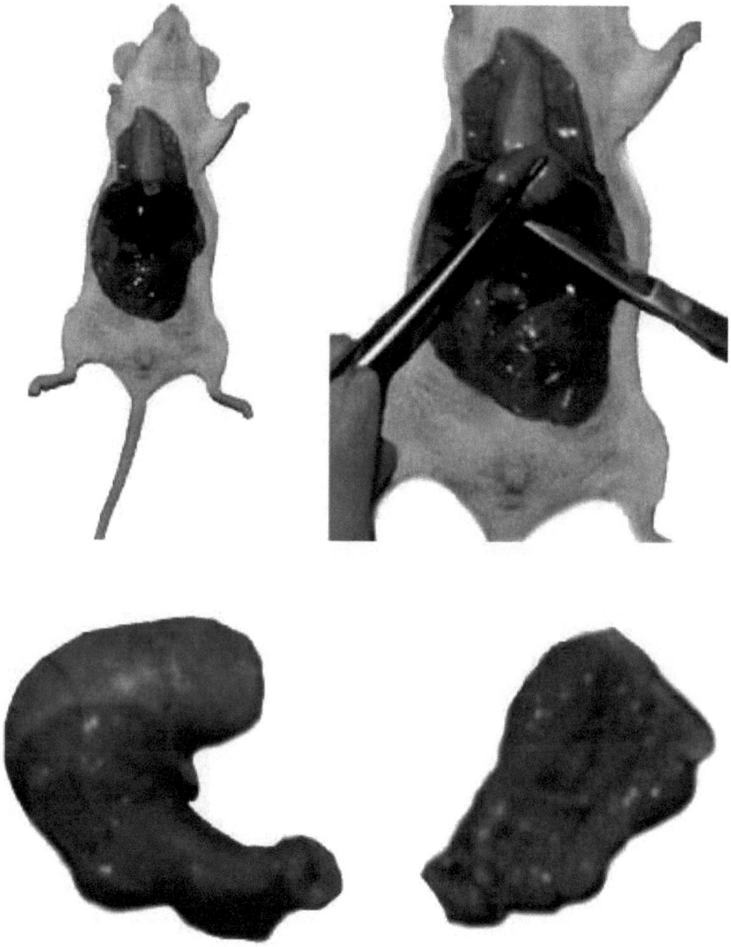

Fig. 2. Collection of mouse stomachs. (**a**) Euthanasia of mouse and opening of abdominal cavity. (**b**) Stomach is moved to the exterior of the body (magnification 2×). (**c**) Stomach is detached from the rest of the gastrointestinal tract and any attached mesentery removed (magnification 8×). (**d**) Stomach is opened along the lesser curvature. The non-squamous epithelium and ingested material are removed (magnification 8×).

4. Open the stomach by cutting along the greater curvature (Fig. 3) and remove residual digesta by gentle scraping with the reverse side of a scalpel blade or similar instrument (see Note 17).
5. Wash the stomach in physiological saline or PBS and blot dry on absorbent paper toweling.
6. The nonglandular region or forestomach (white in appearance) can then be removed (Fig. 3).
7. Flatten the stomach and bisect into two identical halves consisting of both antral and body regions; each half of the stomach can be subjected to a different analysis, as described below (see Notes 17 and 18).

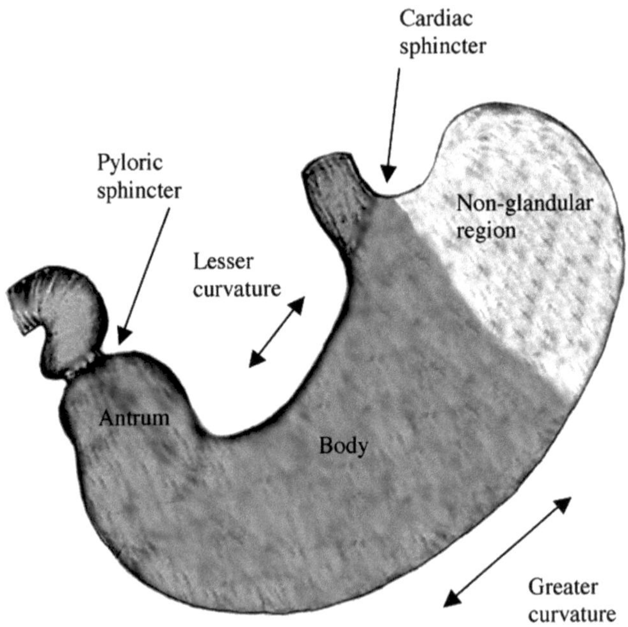

Fig. 3. Anatomy of the murine stomach. The mouse stomach is composed of three major regions, comprising (from proximal to distal): the nonglandular region, the body (or corpus), and the antrum.

3.4.2. Bacteria Culture (see Note 19)

1. Homogenize a half or whole section of stomach from each euthanized mouse (as previously described) in BHI, PBS or similar medium, using either an Ultra Turrax (Janke & Kunkel, IKA) homogenizer or manually, using autoclavable polypropylene pestles and sterile 1.5/2.0 mL tubes.
2. The containers used for homogenization should be tared and then weighed after addition of each tissue piece. In the case of mechanical homogenization, it is important that the Ultra Turrax probe is rinsed with distilled water and ethanol in between each sample.
3. Serially diluted tissue homogenates and culture as described previously.

3.4.3. Urease Assay

1. Add urease reagent in 500 µl or 150 µl aliquots to 1.5 mL tubes or flat-bottomed microtitre plates (Falcon), respectively.
2. Place gastric tissue fragments into the urease reagent (see Note 20) and seal tubes or wells (the latter using clear adhesive tape).
3. Incubate tubes or microtitre plates at room temperature. In the event of *Helicobacter* infection, the color of the urease reagent will change from an orange or yellow color (depending

on the reagent; see Subheading 2.4, item 1) to a deep red, due to urease breakdown of urea into NH_4^+ ions and CO_2. Typically, this should take no longer than 6 h. The color change in samples may be assessed "by eye" or by spectrophotometric analysis (at 550 nm) (17). In the former case, results are recorded as "plus" or "minus," or may be semi-quantified by scoring from 0 to 3+, where a 0 or 1 corresponds to reduced or possibly no infection.

3.4.4. Histology

1. Stomach tissue halves may be prepared for histology by immersion in formalin or, if immunohistochemistry is required, in CRYO-OCT or equivalent cryopreservation medium (see Subheading 2.4, item 2 and 3; Note 21).
2. Tissues should be processed and embedded accordingly. Half stomach pieces should be further dissected into small strips (several mm thick) and embedded such that longitudinal sections containing both antrum and corpus are generated in which mucosa, submucosa and muscularis regions of the tissue are apparent in each strip.
3. To detect *H. pylori* bacteria, which have a relatively small size, it is usually necessary to perform a specialist stain such as the Warthin-Starry or similar silver nitrate-based technique. Alternatively, the bacterium can be detected by immunohistochemistry, using an *H. pylori*-specific antibody. In the case of *H. felis*, which is much larger in size, it is also possible to use a Giemsa stain (16).

3.4.5. Polymerase Chain Reaction

The presence of gastric *Helicobacter* spp. (3) and/or of related enterohepatic species (e.g., *Helicobacter bilis* and *Helicobacter hepaticus*; see Note 1) can be determined by polymerase chain reaction (PCR).

1. Extract DNA from the homogenized stomach using a QIAamp tissue DNA kit (Qiagen), as per manufacturer's instructions (see Note 22).
2. Perform *Helicobacter* genus PCR which targets a 375 bp fragment of the 16S rRNA gene forward and reverse primers: 5′ TAT GAC GGG TAT CCG GC 3′ and 5′ ATT CCA CTT ACC TCT CCC A 3′, respectively. 1 μL of 50 μM of each primer is mixed with 45 μL of PCR Supermix (Invitrogen) and 5 μL DNA.
3. Heat mixtures to 94°C for 2 min, followed by 35 cycles of 94°C for 2 s, 53°C for 2 s and 72°C for 30 s, before holding at 4°C.
4. To test intestine for enteric *Helicobacter* or to test stool (fecal extraction kit (QIAamp stool DNA kit, Qiagen)) can be used in this protocol (see Note 22).

3.4.6. Serology (see Chapter 19 by Dr. Whary)

3.4.7. Assessing Inflammatory and Immune Responses in Mice (see Chapters 21 and 22 by Dr. Rogers)

3.4.8. Epithelial Cell Proliferation

1. BrdU (50 mg/kg) is given intraperitoneal 1 h prior to euthanasia (5).
2. The tissues are then analyzed by immunocytochemistry using anti BrdU Ab (18).
3. Anti PCNA antibody can also be used. The amount of BrdU incorporated into cells correlates with the amount of cell proliferation (11).

3.4.9. Cytokine and vvvvChemokine Response (see Chapters 23 and 24 by Dr. Kurt-Jones)

4. Notes

1. Most studies typically use female mice because these are more tractable and pose fewer problems from an animal husbandry perspective than male animals. Nevertheless, it is important to note that the gender of the animals can have an impact on the experimental results in certain models (19, 20). Mice on a C57BL/6 genetic background have traditionally been employed in *Helicobacter* infection models. For practical reasons, however, this may not always be possible and so alternative mouse backgrounds may be used, e.g., BALB/c, SJL, 129. Again, as for gender, the experimenter needs to be aware that host genetic background can have an impact on the type and severity of the inflammatory disease in animals (2, 21, 22).

   Mice with an SPF status are suitable for *Helicobacter* infection studies. Studies have also reported the use of either gnotobiotic (where the flora is defined), germ-free (or axenic), or even conventional animals. In any case, mice must be free of any enteric *Helicobacter* spp., particularly but not exclusively the two *bona fide* mouse pathogens, *Helicobacter bilis* and *Helicobacter hepaticus*, because these species have been shown to influence the immune responses induced by gastric *Helicobacter* infection (23)

2. Each of the listed *H. pylori* isolates have different characteristics that need to be considered when designing animal experiments, whether these are to investigate chronic infection/gastric carcinogenesis, or the role of cagPAI or other putative virulence factors in inflammation and/or colonization. *H. pylori* SS1 colonizes mice to very high bacterial loads and has an intact cagPAI but its T4SS is nonfunctional. In contrast, the B128 and B128 7.13 strains are cagPAI-positive strains and have functional T4SSs. In our hands (RF and PS, unpublished data), *H. pylori* B128 7.13 typically colonizes mice to very low levels, equivalent to approximately 2 log CFUs lower than that for *H. pylori* SS1. *H. pylori* X47-2AL is cagPAI-negative, T4SS-negative, and colonizes mice to levels comparable to those of *H. pylori* SS1.
3. Alternative agar bases for the growth of *H. pylori* on solid medium include: Columbia Blood Agar Base, Brucella Agar, and Brain-Heart Infusion Agar. Ensure that all glassware has been rinsed free of any traces of detergent as this may affect growth.
4. Skirrow's original recipe has been modified by the addition of an antifungal, amphotericin B (also known as fungizone). Alternatively, *H. pylori* strains can be grown in media to which has been added Dent's Selective Supplement, containing (per liter): 10.0 mg, vancomycin sulfate salt; 5.0 mg, Trimethoprim; 5.0 mg, Cefsulodin; and 5.0 mg Amphotericin B. *H. felis* cannot be grown on medium containing this supplement (RF, unpublished data). Although bacteriologists normally eschew the addition of antibiotics to media used for the routine growth of bacterial species, this is indispensable for the sure growth of helicobacters free of contamination.
5. For optimal growth of *Helicobacter* spp., it is important to use moist plates. The only exception to this is if single colonies are desired.
6. Ensure that any glassware used in the preparation of BHI broth has been briefly rinsed in distilled water to remove any residual detergent prior to use.
7. Heat inactivation of FCS. It is important to remove any complement from FCS because this has been shown to have bactericidal effects against *H. pylori*. For this, the serum needs to be heated to 60°C for a full 30 min. The heat-inactivated serum can then be divided into aliquots and stored at −20 or −80°C.
8. Bacitracin and Nalidixic acid reduces the risk of overgrowth from normal flora present in the mouse stomach. These antibiotics should, however, be omitted if isolating *H. felis* from mouse gastric tissues (RF, unpublished data).
9. *H. felis* in particular, normally grows as "swarms" on plates, akin to swarming *Proteus* spp., however, it can sometimes form

colonies if resuspended in cell culture media, such as RPMI or DMEM, prior to spreading on plates (Invitrogen; RF, unpublished data).

10. For general subculturing purposes, it is advisable to avoid incubating plates for more than 2 days. Nevertheless, bacterial strains that have been isolated from mice, may grow slower and therefore need more time to grow, particularly when coming directly from mouse biopsy samples. In which case, these isolates will need 3–4 days' incubation. Plates for viable number determinations (see below) should be incubated for 3–5 days. *N. B.* All incubation times above are valid for Oxoid anaerobe jars (Anaerojar, AG0025; or Anaerobic jar, HP0011) with the corresponding gas packs (CampyGen 2.5 L, CN0025; or CampyGen 3.5 L, CN0035, respectively), but may vary with other methods of incubation.

    To ensure that bacteria are not excessively passaged in vitro, and therefore do not lose their infectivity or virulence for mice, it is advisable that aliquots of low-passaged isolates be stored at regular intervals and also that excessive subculturing of isolates is avoided

11. Pre-dry blood plates by either:
    (a) Pouring the plates the day before they are required and leaving the plates on the bench overnight.
    (b) Placing the plates with the agar and lid surface facing down for approximately 10 min in the 37°C room/incubator.
    (c) Placing the plates in a biohazard cabinet (with fan on), with the agar and lid surfaces both facing up for approximately 10 min.

    N. B. If plates are to be used to count bacterial loads in mouse gastric biopsies, then ensure that the standard blood agar medium, containing modified Skirrow's Selective Supplement, has been supplemented with bacitracin and nalidixic acid.

12. It is critical to perform a viable count on *H. pylori* suspensions after inoculation of animals. *N. B.* In contrast to the situation for *Escherichia coli* and many standard Gram-negative bacteria, the bacterial densities of *H. pylori* and *H. felis* suspensions cannot be estimated from the A_{600} readings of standardized *E. coli* suspensions because, firstly, *Helicobacter* spp. have very different morphologies and so absorb light differently to *E. coli* and, secondly, it is possible to have turbid *Helicobacter* cultures containing predominantly nonviable, coccoid forms. Thus, the measurement of *Helicobacter* numbers by A_{600} measurement is erroneous and should not be used.

13. As a gavage, the authors recommend using 1.0 mL disposable tuberculin syringes, fitted with 23 gauge needles, to which have

been affixed small sections (6–8 cm long) of disposable teflon or polycarbonate catheters (internal diameter = 0.58 mm).

14. Although many researchers still perform multiple inoculations (typically three intragastric gavages over a period of 5 days), this is not only unnecessary but causes undue stress in the animals and should therefore be avoided at all costs.
15. Although this treatment induced neoplasms in the mouth and nonglandular forestomach of animals, no neoplastic lesions were induced in the glandular region of the stomach, possibly due to the strain of mouse used (9).
16. High levels of dietary salt enhance *H. pylori* colonization and cause increased proliferation in the proximal body and antrum with a multifocal reduction in parietal cell numbers in the proximal body, resulting in the elongation of gastric pits (11). Nevertheless, this treatment does not seem to induce neoplasia in mice (6, 11).
17. An alternative method for dissection of the stomach involves first the removal of the nonglandular tissue by cutting in a diagonal direction across the intact stomach. The stomach can then be opened by cutting along the lesser curvature and treated as above.
18. The authors strongly advise that, ideally, stomachs should be dissected into no more than two fragments as, otherwise, this can lead to sampling errors due to the intrinsic differences between the various regions of the stomach and known patterns of colonization in the stomach.
19. It is important to note that changes in gastric tissue mass, due to severe inflammation, may introduce a bias in the data if CFU are calculated per gram, rather than per stomach (24). Nevertheless, this is unlikely to have a significant effect if mice are age- and gender-matched, thus either type of calculation method is possible. Bacteriological detection of *Helicobacter* infection remains the "gold standard" method for detecting *H. pylori* infection, but works less well for *H. felis* because, as indicated above, this species does not normally form isolated colonies. This method also allows the determination of bacterial numbers or loads in the stomach, expressed as either CFU/mg gastric tissue or CFU/whole stomach.
20. This assay typically works best for small biopsies, rather than half sections of stomach.
21. To ensure the orientation of tissues for classical histology, half sections of stomach should be laid flat and pinned on firm supports (e.g., cardboard or foam), prior to immersion in formalin. Alternatively, tissues can be plunged quickly in formalin and then flattened against the walls of plastic specimen containers and 1–2 h later, dropped into the formalin within the containers.

22. PCR results can be influenced by the presence of inhibitors and/or potential problems of cross-contamination, thus suitable positive and negative controls must always be included in these assays.

Acknowledgments

RF and PS acknowledge grant support (APP1006010 and APP1002947) from the National Health and Medical Research Council (NHMRC), Australia. The work performed at MIMR is supported by the Victorian Government's Operational Infrastructure Support Program. RF and PS are NHMRC Senior Research Fellows.

References

1. Danon SJ, Eaton KA (1998) The role of gastric *Helicobacter* and N-methyl-N′-nitro-N-nitrosoguanidine in carcinogenesis of mice. Helicobacter 3:260–268
2. Sakagami T, Dixon M, O'Rourke J, Howlett R, Alderuccio F, Vella J et al (1996) Atrophic gastric changes in both *Helicobacter felis* and *Helicobacter pylori* infected mice are host dependent and separate from antral gastritis. Gut 39:639–648
3. Cai X, Carlson J, Stoicov C, Li H, Wang TC, Houghton J (2005) *Helicobacter felis* eradication restores normal architecture and inhibits gastric cancer progression in C57BL/6 mice. Gastroenterology 128:1937–1952
4. Enno A, O'Rourke JL, Howlett CR, Jack A, Dixon MF, Lee A (1995) MALToma-like lesions in the murine gastric mucosa after long-term infection with *Helicobacter felis* A mouse model of *Helicobacter pylori*-induced gastric lymphoma. Am J Pathol 147:217–222
5. Ferrero RL, Thiberge JM, Huerre M, Labigne A (1998) Immune responses of specific-pathogen-free mice to chronic *Helicobacter pylori* (strain SS1) infection. Infect Immun 66:1349–1355
6. Rogers AB, Taylor NS, Whary MT, Stefanich ED, Wang TC, Fox JG (2005) *Helicobacter pylori* but not high salt induces gastric intraepithelial neoplasia in B6129 mice. Cancer Res 65: 10709–10715
7. Tu S, Bhagat G, Cui G, Takaishi S, Kurt-Jones EA, Rickman B et al (2008) Overexpression of interleukin-1β induces gastric inflammation and cancer and mobilizes myeloid-derived suppressor cells in mice. Cancer Cell 14:408–419
8. Wang TC, Dangler CA, Chen D, Goldenring JR, Koh T, Raychowdhury R et al (2000) Synergistic interaction between hypergastrinemia and *Helicobacter* infection in a mouse model of gastric cancer. Gastroenterology 118:36–47
9. Han SU, Kim YB, Joo HJ, Hahm KB, Lee WH, Cho YK et al (2002) *Helicobacter pylori* infection promotes gastric carcinogenesis in a mice model. J Gastroenterol Hepatol 17:253–261
10. Shimizu N, Kaminishi M, Tatematsu M, Tsuji E, Yoshikawa A, Yamaguchi H et al (1998) *Helicobacter pylori* promotes development of pepsinogen-altered pyloric glands, a preneoplastic lesion of glandular stomach of BALB/c mice pretreated with N-methyl-N-nitrosourea. Cancer Lett 123:63–69
11. Fox JG, Dangler CA, Taylor NS, King A, Koh TJ, Wang TC (1999) High-salt diet induces gastric epithelial hyperplasia and parietal cell loss, and enhances *Helicobacter pylori* colonization in C57BL/6 mice. Cancer Res 59:4823–4828
12. Lee A, O'Rourke J, De Ungria MC, Robertson B, Daskalopoulos G, Dixon MF (1997) A standardized mouse model of *Helicobacter pylori* infection: Introducing the Sydney strain. Gastroenterology 112:1386–1397
13. Israel DA, Salama N, Arnold CN, Moss SF, Ando T, Wirth HP et al (2001) *Helicobacter pylori* strain-specific differences in genetic content, identified by microarray, influence host inflammatory responses. J Clin Invest 107:611–620
14. Fox JG, Batchelder M, Marini R, Yan L, Handt L, Li X et al (1995) *Helicobacter pylori*-induced gastritis in the domestic cat. Infect Immun 63:2674–2681

15. Lee A, Hazell SL, O'Rourke J, Kouprach S (1988) Isolation of a spiral-shaped bacterium from the cat stomach. Infect Immun 56:2843–2850
16. Ferrero RL, Thiberge JM, Kansau I, Wuscher N, Huerre M, Labigne A (1995) The GroES homolog of *Helicobacter pylori* confers protective immunity against mucosal infection in mice. Proc Natl Acad Sci U S A 92:6499–6503
17. Corthesy-Theulaz I, Porta N, Glauser M, Saraga E, Vaney AC, Haas R et al (1995) Oral immunization with *Helicobacter pylori* urease B subunit as a treatment against *Helicobacter* infection in mice. Gastroenterology 109:115–121
18. Fox JG, Li X, Cahill RJ, Andrutis K, Rustgi AK, Odze R et al (1996) Hypertrophic gastropathy in *Helicobacter felis*-infected wild-type C57BL/6 mice and p53 hemizygous transgenic mice. Gastroenterology 110:155–166
19. Fox JG, Rogers AB, Ihrig M, Taylor NS, Whary MT, Dockray G et al (2003) *Helicobacter pylori*-associated gastric cancer in INS-GAS mice is gender specific. Cancer Res 63:942–950
20. Fox JG, Wang TC, Rogers AB, Poutahidis T, Ge Z, Taylor N et al (2003) Host and microbial constituents influence *Helicobacter pylori*-induced cancer in a murine model of hypergastrinemia. Gastroenterology 124: 1879–1890
21. De Bock M, Decostere A, Van den Bulck K, Baele M, Duchateau L, Haesebrouck F et al (2005) The inflammatory response in the mouse stomach to *Helicobacter bizzozeronii*, *Helicobacter salomonis* and two *Helicobacter felis* strains. J Comp Pathol 133:83–91
22. Kim JS, Chang JH, Chung SI, Yum JS (2001) Importance of the host genetic background on immune responses to *Helicobacter pylori* infection and therapeutic vaccine efficacy. FEMS Immunol Med Microbiol 31:41–46
23. Lemke LB, Ge Z, Whary MT, Feng Y, Rogers AB, Muthupalani S et al (2009) Concurrent *Helicobacter bilis* infection in C57BL/6 mice attenuates proinflammatory *H. pylori*-induced gastric pathology. Infect Immun 77:2147–2158
24. Sutton P (2007) Considering increased mouse stomach mass when calculating prophylactic vaccine efficacy against *Helicobacter pylori*. Helicobacter 12:210–212

Chapter 21

Gastric *Helicobacter* spp. in Animal Models: Pathogenesis and Modulation by Extragastric Coinfections

Arlin B. Rogers

Abstract

Animal models are used to study complex host, microbial, and environmental influences associated with gastric *Helicobacter* infection. Evidence that gastric helicobacters are pathogenic in animals first came from ferrets. Felids, nonhuman primates, and many other species also harbor stomach helicobacters. Today, mice are preferred by most researchers for scientific investigation because of cost-efficiencies, rapid reproduction, choice of laboratory reagents, and availability of genetically engineered models. Infection with *Helicobacter felis* or *H. pylori* Sydney strain-1 in appropriate mouse strains produces disease with remarkable similarities to *H. pylori* in humans. Due to recent advances in genetic engineering, in vivo imaging, and system-wide genomics and proteomics, these models will become even more widespread in the future. Recently, it has been shown that extragastric infections can dramatically affect the severity of disease induced by gastric *Helicobacter* spp. through heterologous immunity. These models provide proof-of-principle for the "African enigma" wherein gastric cancer is underrepresented in low-lying tropical countries with concurrently high *H. pylori* and internal parasite prevalence. Helicobacter gastritis and carcinogenesis in mouse models may be augmented or ameliorated by other infectious agents depending on the character of the invoked immune response. Knowledge gained from the Human Microbiome Project and other investigations is certain to shed new light on the influence of extragastric bacterial, viral, fungal, and parasitic coinfections on *H. pylori*-associated peptic ulcer disease and gastric adenocarcinoma.

Key words: *Helicobacter* infections, *Helicobacter pylori*, Mucosal immunity, Stomach neoplasms, Disease models, Animal

1. Introduction

Animal models are an indispensable resource for investigating the causes of gastric cancer and for identifying new preventative and therapeutic measures. In the late 1980s, experimental *Helicobacter mustelae* infection of domestic ferrets provided the first proof-of-principle that helicobacter colonization was sufficient to incite chronic gastritis,

JeanMarie Houghton (ed.), *Helicobacter Species: Methods and Protocols*, Methods in Molecular Biology, vol. 921,
DOI 10.1007/978-1-62703-005-2_21, © Springer Science+Business Media, LLC 2012

helping to dispel skepticism over the pathogenic role of *H. pylori* in humans (1). Today, mice have emerged as the model of choice for most laboratories. These models have demonstrated immunomodulation between gastric and lower bowel *Helicobacter* spp., and highlighted the importance of both intestinal and extraintestinal coinfections when interpreting phenotypic outcomes such as inflammation severity and tumor development (2–4). Coinfections almost certainly affect the development of *H. pylori*-associated disease in human populations as well, as evidenced by marked regional differences in stomach cancer incidence (5). This chapter provides a review of gastric helicobacters in animal models with an emphasis on the mouse, and considers the importance of immunomodulation by coinfection with parasites and/or gut microbiota, including enterohepatic *Helicobacter* spp. (EHS).

2. Gastric *Helicobacter* Infection in Animal Models

Indigenous *Helicobacter* spp. in animals may invoke gastritis and cancer just as *H. pylori* does in humans. Gastric *Helicobacter* spp. have been recovered from rats, dogs, pigs, nonhuman primates, and many wild species including birds and aquatic mammals (6–13). As described below, ferrets, nonhuman primates, and cats (both domestic and sylvan) have received some attention as models of the human disease. Nevertheless, rodents are by far the most utilized animal models. Today the two most commonly used animal species for experimental infectious gastric carcinogenesis studies are the mouse and Mongolian gerbil. In this chapter gastric helicobacters in larger animal species are reviewed, followed by a more exhaustive comparison of rodent models. Finally, the importance of immunomodulatory effects from coinfections will be discussed, including EHS that are ubiquitous in research colonies (14).

2.1. Gastric *Helicobacter* spp. in Non-Rodent Species

2.1.1. Helicobacter Mustelae in Ferrets

The ferret was the first animal identified with a persistent gastric *Campylobacter*-like organism (later *H. mustelae*) that mirrored *H. pylori* infection of humans (15). Until mouse models emerged, this was the most popular animal system for studying gastric *Helicobacter* infection and it remains in use to this day (1, 16, 17). *H. mustelae*, a natural pathogen of ferrets, has many of the same biochemical, molecular, and disease-inducing characteristics as *H. pylori*. It is urease positive, motile, and binds to similar receptors (18, 19). Like *H. pylori*, *H. mustelae* persistently infects the inflamed mucosa, with colonization occurring shortly after weaning (18). Experimental inoculation of *H. mustelae* into naive ferrets induces a chronic gastritis identical in character to that of naturally infected animals. Moreover, the ferret stomach closely resembles the human stomach in anatomic and physiologic features (15). Ferrets secrete gastric acids and proteolytic enzymes under basal conditions; like humans with *H. pylori*, *H. mustelae*-infected ferrets frequently develop hypergastrinemia (20). Adenocarcinoma of the antrum and pylorus is linked to *H. mustelae* infection in association with upregulated cell replication (21, 22). *H. mustelae*-infected ferrets may develop either diffuse antral gastritis or multifocal atrophic gastritis equivalent to the human disease (1, 23). Thus, *H. mustelae* infection of ferrets represents an excellent model to study the pathogenesis of *H. pylori*-associated gastritis and cancer. Limitations of the ferret model include the relatively high cost to purchase and maintain animals, long disease course, and lack of commercial reagents available. Nevertheless, the model remains useful for the evaluation of therapeutic interventions due to the coadapted nature of host and pathogen.

2.1.2. Gastric Helicobacter spp. in Cats

Domestic cats may be infected naturally and experimentally with a number of potentially pathogenic gastric *Helicobacter* spp. (24, 25). Indeed, some *Helicobacter* spp. colonize both the human and feline stomach, raising the possibility of interspecies transmission (26–28). One outbreak of *H. pylori*-associated gastritis among cats in a commercial breeding facility was attributed to human-to-animal (anthropozoonotic) transmission (29). Experimental *H. pylori* inoculation results in pangastric colonization of the feline stomach with antral inflammation resembling the human disease (30–33). Chronic *H. pylori* infection of cats results in prominent follicular gastritis and/or mucosal hyperplasia and dysplasia (34). Cats were identified as the natural source of *H. felis* which is used widely in murine experimental models (35). *H. felis* infection of cats is characterized by antral-predominant gastritis with expansile submucosal lymphoid follicles and a lesser intramucosal component (35, 36). Some investigators postulate that chronic *Helicobacter* infection contributes to the high incidence of gastrointestinal lymphoid neoplasms in the cat (37). In addition to domestic cats, wild felids harbor a variety of gastric helicobacters and, like humans, exhibit widely variable disease presentation. For example, captive cheetahs develop chronic gastritis that is an important cause of mortality (38). But, whereas both wild and captive cheetahs harbor gastric *Helicobacter* spp., disease is limited to captive populations and is not readily attributable solely to helicobacter infection (39, 40).

2.1.3. Gastric Helicobacter spp. in Nonhuman Primates

Nonhuman primates have a high prevalence of endogenous *Helicobacter* spp. infections, including *H. pylori* in rhesus macaques (41, 42). With significant effort, macaque colonies free of *H. pylori* have been created through hand-rearing of newborn monkeys (43). As a result, controlled experiments of the timing, infectious dose, and disease progression of *H. pylori* in macaques are now possible (44, 45). Because of the closely related genetic profiles between macaques and humans, transcriptional profiles following gastric *H. pylori* infection exhibit high homology between species (46). Nevertheless, given the long latent period, further work will be needed to determine whether experimental gastric carcinoma is induced by *H. pylori* in rhesus macaques and other nonhuman primates.

2.2. Gastric Helicobacter spp. in Rodent Models

2.2.1. Gastric Helicobacter spp. in Mongolian Gerbils

Mongolian gerbils can be chronically infected with human isolates of *H. pylori*, and may develop gastroduodenitis, ulcers, and antral cancer closely resembling the human disease (47–49). Gerbils develop both a marked submucosal follicular response combined with moderate to severe diffuse infiltration of the lamina propria with granulocytes and mononuclear cells. Dysplastic glands invade the submucosa early in the disease course, and can sometimes markedly expand this compartment with compression of the subjacent tunica muscularis. Recently, a rapid model of tumorigenesis was identified in gerbils infected with an *H. pylori* B128 variant (strain 7.13) that transactivated β-catenin signaling through a cagA-dependent mechanism (50). In addition to developing cancer, *H. pylori*-infected Mongolian gerbils are unique among rodents in their susceptibility to peptic ulcer disease (51). Unfortunately, interlaboratory differences in disease outcome and interpretation of *H. pylori* infection in Mongolian gerbils are significant, and likely the result of genetic, environmental and microbial circumstances unique to each setting. Unlike inbred strains of mice, Mongolian gerbils typically used in *H. pylori* infection studies are outbred and exhibit significant genetic diversity (52). Of perhaps greater importance, resident microbiota in different animal facilities may greatly influence responses to *H. pylori* infection. Heterologous immunity, both proinflammatory and anti-inflammatory, has been documented in mouse models of coinfection with enteric protozoa and helminths (2, 3). Although not as well characterized as mice, Mongolian gerbils are likely to remain in use as important and highly representational models of *H. pylori* gastric carcinogenesis, and the impact of coinfections on disease progression.

2.2.2. Gastric Helicobacter spp. in Wild-Type Mice

H. felis in WT Mice

Mice can be persistently colonized by a number of gastric *Helicobacter* spp., but the two most commonly used to study gastritis and cancer are *H. felis* and *H. pylori*. *H. felis*, originally isolated from cats, was the first bacterium shown to induce chronic gastritis in experimentally infected mice (53). *H. felis* is a much larger organism than *H. pylori*, with important differentiating ultrastructural and genetic features (54). One limitation of the *H. felis* model is that it lacks the cag pathogenicity-associated island (PAI) found in *H. pylori* (55). Nevertheless, the ability to induce severe gastritis and carcinoma in C57BL/6 and other WT mice makes *H. felis* a highly valuable model for studying inflammation-associated gastric carcinogenesis. Indeed, murine gastritis caused by *H. felis* is more severe than that induced by *H. pylori*, and only the former produces carcinoma in WT C57BL/6 mice (56). As with most mouse models of *Helicobacter* infection, phenotypic outcomes are strain specific. Whereas C57BL/6 mice infected with *H. felis* develop severe gastritis progressing to cancer, outbred Swiss Webster mice acquire only a moderate self-limiting gastritis (57). Two different laboratories inoculated different inbred strains of mice with *H. felis* and tabulated the intensity of inflammation; in BALB/c mice inflammation was minimal, in C3H/He moderate and was most severe in C57BL/6 (58, 59).

H. pylori in WT Mice

H. pylori has a limited natural host range, to date having been recovered only from humans, nonhuman primates and domestic cats (57). Experimentally, *H. pylori* will infect mice, Mongolian gerbils, and a number of domestic animal species (60). Early studies of *H. pylori* in rodents were hampered by poor adapation leading to low colonization kinetics and a weak phenotype. Lee and colleagues in Australia successfully adapted a strain of *H. pylori* to the mouse stomach through serial inoculation. This became known as the Sydney strain-1 (SS1) isolate, which is the most commonly used infecting agent used in mouse studies today (61). Other human strains including B128 also exhibit virulence in certain mouse models (62). Whereas B6 and other strains of mice develop chronic gastritis when infected with *H. pylori* SS1, to date the only WT mice proven to develop tumors are C57BL/6 × 129S6/SvEv (B6129) mice. These mice may display gastric intraepithelial neoplasia (GIN) as early as 15 months of age (63). Longer studies have not yet been performed to determine whether they will progress to invasive adenocarcinoma. Moreover, infection of inbred 129S in an experimental setting has not been reported. These mice often exhibit exaggerated inflammatory responses to gastrointestinal infections, and may prove useful for the investigation of gastric helicobacter infection as well. Even in the absence of cancer induction, mouse models of *H. pylori* offer important molecular insights into cancer determinants, such as the demonstration of oxidative DNA mutations in C57BL/6 Big Blue mice carrying a lambda phage mutation biomarker (64).

2.2.3. Gastric Helicobacter spp. in Genetically Engineered Mice

Gastric Carcinogenesis in INS-GAS Transgenic Mice

The INS-GAS model was the first published report of *H. pylori*-induced gastric carcinoma in mice (62, 65). Hypergastrinemic INS-GAS mice that constitutively express humanized gastrin under the rat insulin promoter acquire spontaneous gastric tumors within 2 years, and develop severe gastritis and carcinoma within months when infected with *H. felis* or *H. pylori* (62, 66). Because of the rapid disease course, putative carcinogenic microbial factors are amenable to study in this system. For example, it was shown that deletion of *H. pylori* cagE (picB) delayed but did not prevent progression to cancer (62). More recently, this model was used to demonstrate that exposure to the Swedish smokeless tobacco snus promoted gastric carcinogenesis (67). A unique and important feature of the INS-GAS model is male-predominant disease expression that mirrors human gender dimorphic cancer risk (65). Studies with INS-GAS mice showed that estrogen has a protective effect against cancer in both males and females infected with *H. pylori* (68). Importantly, as with other genetically engineered mice (GEM), the strain background on which the INS-GAS transgene is expressed influences phenotypic outcome. Whereas INS-GAS mice on the FVB strain background rapidly progress to cancer when infected with *Helicobacter* spp., C57BL/6 mice expressing the same transgene exhibit a markedly attenuated phenotype (69).

Gastric Carcinogenesis in p27(kip1)$^{-/-}$ Mice

In contrast to p53$^{+/-}$ and Apc-haploinsufficient mice, deletion of the cell-cycle kinase inhibitor p27(kip1) in B6 mice conferred susceptibility to gastric carcinoma following *H. pylori* infection (70). This was the first demonstration of *H. pylori*-induced gastric carcinoma in a knockout (as opposed to the transgenic INS-GAS) mouse. Deletion of this tumor-suppressor gene increased susceptibility to cancer at 60 and 75 weeks post-infection versus WT controls, with increased epithelial proliferation and decreased apoptosis in the p27$^{-/-}$ cohort (70). This should prove to be a useful model for the study of *H. pylori* carcinogeneis, and may help spotlight new molecular targets for the interruption of stomach cancer in humans.

Gastric Carcinogenesis in Tff2$^{-/-}$ Mice Provides New Insights into SPEM

Trefoil factor 2 (TFF2; spasmolytic polypeptide), is normally expressed in mucous neck cells of the corpus and in basally located epithelial cells of the antrum (71). In humans, expansion of TFF2^{+} cells associated with *H. pylori*-induced antralization of the fundic mucosa is called spasmolytic polypeptide-enhancing metaplasia (SPEM) (71). The same process is observed in *Helicobacter*-infected mice with pseudopyloric metaplasia (72,73). Pseudopyloric metaplasia is a preneoplastic lesion associated with chronic *Helicobacter* infection (74). In contrast, mucous metaplasia, which is defined by expansion of foamy cells replacing oxyntic glands, may occur in the absence of *Helicobacter* infection (63). Because TFF2 is expressed both in mucous metaplasia and pseudopyloric metaplasia, a diagnosis of SPEM in mice must be qualified by the overall histologic picture (75). Moreover, whereas SPEM in humans was once believed to be a driver of gastric preneoplasia, data from mice prove that TFF2 plays a protective rather than promotional role in carcinogenesis. TFF2 exhibits an immunoregulatory function in vitro, and mice lacking this gene demonstrate more severe inflammation challenged with proinflammatory chemicals or *Helicobacter* infection. (76). Moreover, TFF2$^{-/-}$ mice develop mucous metaplasia and pseudopyloric metaplasia with the same frequency as WT controls, and the knockouts have a significantly increased risk for tumors of the pyloric antrum (75). Taken together, results from mouse models suggest that spasmolytic polypeptide has an antitumorigenic rather than protumorigenic function in gastric carcinogenesis, and that it is dispensable for the development of pseudopyloric metaplasia. Nevertheless, SPEM remains useful as a biomarker of gastric preneoplasia risk in humans infected with *H. pylori* (77).

Paradoxical Tumor Protection in p53 Haploinsufficient Mice

The p53 tumor suppressor gene is a housekeeping gene that is responsive to DNA damage, orchestrates repair, helps regulate the cell cycle and plays a role in transcription (78). Humans heterozygous for p53 have Li-Fraumeni syndrome and are at risk for a variety of tumors. Befitting its role as a central regulator of DNA integrity and cell cycling, double knockout of p53 in mice results in significantly shortened lifespan (79). On the other hand, $p53^{+/-}$ heterozygotes are viable and susceptible to a variety of spontaneous tumors; however, none have been recorded in the glandular stomach (80). A study of chronic *H. felis* infection in p53 hemizygous mice found that after one year both WT and p53 hemizygous mice showed severe adenomatous and cystic hyperplasia of the gastric surface foveolar epithelium. The proliferation markers BrdU and PCNA were markedly increased in both groups of infected mice. Not unexpectedly, p53 hemizygous mice had a higher proliferative index than WT mice, although this fell short of statistical significance (81). In spite of increased cell turnover, in a separate study p53 haploinsufficiency resulted in parodoxical protection against tumor development that was attributed to depressed Th1 immune responses in the $p53^{+/-}$ cohort (56). The development of frank neoplasia in this mouse model may require either additional mutations, genetic events, or a longer time period of infection. The p53 heterozygous *H. felis* infected mouse may also be useful for studies on cocarcinogenesis, given the increased sensitivity that $p53^{+/-}$ mice have to tumor induction by a variety of chemicals (82).

Mixed Cancer Phenotype in Apc Haploinsufficient Mice

Mutation of the adenomatosis polyposis coli (APC) gene is indisputably associated with familial adenomatosis polyposis (FAP) and increased risk of colon cancer in humans. However, the syndrome thus far has been linked to increased risk of gastric cancer only in Japan (83). This may reflect different population exposures to both infectious and environmental gastric carcinogens. There are two well-described mouse models of Apc deficiency. The first, and most widely adopted, is the $Apc^{Min/+}$ mouse. This mouse was produced from ethylnitrosuria-induced random mutagenesis. The phenotype of widely disseminated intestinal adenomas was discovered following unexpectedly early mortality of mice due to intestinal hemorrhage (84). Whereas early reports limited proliferative lesions to the intestinal tract, a more recent study reported spontaneous gastric tumors in $Apc^{Min/+}$mice (83). A second mouse model of Apc deficiency, developed through targeted gene deletion, was designated Apc1638 (85). The phenotype of Apc1638 mice is similar to that of $Apc^{Min/+}$mice. *Apc*1638 mice were infected with *H. felis* to determine whether the mutation increased risk of gastric cancer following infection with a tumor-promoting bacterium. Interestingly, infected *Apc*1638 mice had less proliferation and inflammation in the corpus than WT mice, lower anti-*H. felis* serum titers, and higher bacterial colonization and gastric urease production compared with WT mice (86). The *Apc*1638 truncating mutation did lead to spontaneous gastric dysplasia and polyposis of the antrum and pylorus, but this was not worsened by *H. felis* infection. Like p53, *Apc* haploinsufficiency appears to downmodulate immune responses, demonstrating the complex interactions between microbes and different host signaling pathways.

3. Modulation of Gastric Helicobacter Disease by Coinfections

Whereas >50% of the world's population is infected by *H. pylori*, only a small percentage develop peptic ulcer disease or gastric adenocarcinoma (87). Disease expression is determined by numerous factors including age at the time of infection, virulence of the isolate, sex (males have a higher risk of cancer), and host genetics (5, 88). Recently, it has become clear that another major variable is coinfection with other bacteria, viruses, and eukaryotic parasites. Type 2 immunity to parasitic infection has been invoked as a plausible explanation for the "African enigma," wherein rates of gastric cancer in tropical regions are below expected given the very high prevalence of *H. pylori*, which is dependent on Th1-type immunity for disease progression (89). Moreover, coinfection with a Th1-inducing agent may accelerate *H. pylori* disease (3). In this section data from mouse models are presented to directly support the hypothesis that heterologous immunity from extragastric infections may significantly influence gastric disease outcomes in individuals with *H. pylori* infection.

3.1. Disease-Amplifying Coinfections in Mice

Generally speaking, infectious agents that provoke a Th1-type immune response, regardless of tissue site, have the potential to augment disease induced by gastric *Helicobacter* spp. The most direct example of this was reported by Stoicov et al, who showed that infection of BALB/c mice with *Toxoplasma gondii*, a ubiquitous protozoan parasite, polarized mucosal immune responses from the usual Th2-type to a Th1-type (3). This resulted in a significant increase in gastritis and preneoplastic progression in the stomach of mice infected with *H. felis*. At the systemic level, *T. gondii* altered humoral immunity by decreasing relative levels of anti-*H. felis* type II antibody (IgG1) and increasing type I (IgG2a). Simultaneously, *T. gondii* increased expression of type I cytokines in the *H. felis*-infected gastric mucosa including interferon-γ, IL-12, and IL-1β while decreasing the Th2 cytokines IL-4 and IL-10. In agreement with other animal models and the human disease, polarized type I immune responses in the stomach decreased bacterial colonization levels while increasing overt inflammation and epithelial alterations including tumor-associated dysplasia (90). This agrees with other work showing that proinflammatory gut microbes such as *H. hepaticus* can promote extraintestinal cancers including mammary and liver carcinomas (91, 92). Whereas it remains to be confirmed that Th1-invoking coinfections worsen *H. pylori* disease in humans, it is interesting to note that smoking, which is known to produce systemic immune activation and oxidative stress, increases the risk of stomach cancer (93).

3.2. The African Enigma

Whereas invocation of a polarized type I immune response by extragastric coinfection can worsen helicobacter gastritis, systemic polarization to a type II response may ameliorate disease. Associative connections between environment and *H. pylori* disease have been made by Pelayo Correa and others, who note that tropical regions endemic for *H. pylori* exhibit lower-than-expected rates of gastric adenocarcinoma, whereas temperate and mountain regions with the same infection prevalence show a very high incidence of cancer (89, 94). Infection with *Schistosoma mansoni* has been proposed as one factor in the protection against helicobacter gastritis in low-lying tropical regions (95). The conundrum was originally known as the "African Enigma" based on the curiously low incidence of stomach cancer in sub-Saharan countries with high internal parasite burdens (96). Some observers attributed the phenomenon to shorter overall lifespans, whereas others dismissed the concept entirely (97).

3.3. Disease-Ameliorating Coinfections in Mice

Direct laboratory evidence that intestinal helminth infection can decrease the severity of helicobacter-induced gastric disease was reported in a 2000 paper by Fox et al using a mouse model of coinfection with *H. felis* and the murine nematode *Heligmosomoides polygyrus* (2). The authors found that *H. polygyrus* infection increased type II and decreased type I humoral immunity to *H. felis*, in association with increased bacterial colonization of the stomach. Histologic markers of epithelial damage all were decreased in the coinfected versus *H. felis*-noninfected mice although, interestingly, there was no discernible difference in the level of leukocyte infiltration of the stomach between groups. Nevertheless, preneoplastic dysplasia was significantly attenuated in the mice infected with the intestinal parasite, in support of the heterologous immunity argument for the African Enigma. Subsequently, Lemke et al showed that mice infected with the EHS organism *H. bilis* in the lower bowel were protected from *H. pylori*-induced gastritis and dysplasia.(4) In this model, both inflammation and epithelial changes were muted in the stomachs of *H. pylori*-infected mice harboring *H. bilis.* Interestingly, whereas the major type I cytokines including interferon-γ, TNF-α, and IL-1B all were reduced in the stomachs of mice coinfected with *H. bilis*, there was a less-robust amplification of anti-inflammatory markers such as IL-10, IL-13 and TGF-β. Thus, whereas *H. bilis* reduced *H. pylori*-associated disease in a similar manner to the effect of *H. polygyrus* in mice infected with *H. felis*, the former model appeared less reliant on induction of a Th2-type immune response to dampen the gastric disease. Therefore, it seems likely that multiple pathways come into play in the complex crosstalk between gastric helicobacter disease and extragastric coinfections, and that both host and infectious agent factors contribute to the ultimate phenotype that emerges.

4. Summary

Mouse models of gastric helicobacter infection are highly faithful to *H. pylori* infection of humans in histologic presentation and molecular alterations (74, 90). These models have validated the tumorigenic potential of chronic helicobacter colonization and provided insights into host–pathogen relationships that mold disease outcomes. Recently, it has been shown that extragastric infections can have a strong influence on the progression of gastritis and cancer induced by gastric helicobacters. Such infections may augment or ameliorate gastric disease depending on location and the type of immune response elicited. These findings are consistent with the high degree of geographic variation in *H. pylori* disease outcomes in human populations depending on the nature of endemic coinfections (94). In mice, helminth and protozoal parasites have been shown to impact the progression of gastritis and preneoplasia induced by *H. felis*, and EHS in the lower bowel impact the phenotype of disease invoked by *H. pylori* (2–4). Emerging recognition of other background immunomodulating infections such as murine norovirus will further our understanding of how heterologous mucosal immunity modulates chronic inflammation and experimental tumorigenesis both within and outside the gastrointestinal tract (98). Animal models will play an important role in these investigations, and will continue to provide insights into how complex host–pathogen networks influence likelihood of clinical disease in humans infected with *H. pylori*, and how these networks might be therapeutically modulated to improve patient outcomes.

References

1. Fox JG, Correa P, Taylor NS et al (1990) Helicobacter mustelae-associated gastritis in ferrets. An animal model of Helicobacter pylori gastritis in humans. Gastroenterology 99:352–361
2. Fox JG, Beck P, Dangler CA et al (2000) Concurrent enteric helminth infection modulates inflammation and gastric immune responses and reduces helicobacter-induced gastric atrophy. Nat Med 6:536–542
3. Stoicov C, Whary M, Rogers AB et al (2004) Coinfection modulates inflammatory responses and clinical outcome of Helicobacter felis and Toxoplasma gondii infections. J Immunol 173:3329–3336
4. Lemke LB, Ge Z, Whary MT et al (2009) Concurrent Helicobacter bilis infection in C57BL/6 mice attenuates proinflammatory H. pylori-induced gastric pathology. Infect Immun 77:2147–2158
5. Wroblewski LE, Peek RM Jr, Wilson KT (2010) Helicobacter pylori and gastric cancer: factors that modulate disease risk. Clin Microbiol Rev 23:713–739
6. Whary MT, Fox JG (2004) Natural and experimental Helicobacter infections. Comp Med 54:128–158
7. Van den Bulck K, Decostere A, Baele M et al (2005) Identification of non-Helicobacter pylori spiral organisms in gastric samples from humans, dogs, and cats. J Clin Microbiol 43:2256–2260
8. Seymour C, Lewis RG, Kim M et al (1994) Isolation of Helicobacter strains from wild bird and swine feces. Appl Environ Microbiol 60:1025–1028
9. Fox JG, Taylor NS, Howe S et al (2006) Helicobacter anseris sp. nov. and Helicobacter brantae sp. nov., isolated from feces of resident Canada geese in the greater Boston area. Appl Environ Microbiol 72:4633–4637
10. Harper CG, Feng Y, Xu S et al (2002) Helicobacter cetorum sp. nov., a urease-positive Helicobacter species isolated from dolphins and whales. J Clin Microbiol 40: 4536–4543
11. Harper CG, Xu S, Rogers AB et al (2003) Isolation and characterization of novel Helicobacter spp. from the gastric mucosa of harp seals Phoca groenlandica. Dis Aquat Organ 57:1–9
12. Harper CM, Dangler CA, Xu S et al (2000) Isolation and characterization of a Helicobacter sp. from the gastric mucosa of dolphins, Lagenorhynchus acutus and Delphinus delphis. Appl Environ Microbiol 66:4751–4757
13. Harper CM, Xu S, Feng Y et al (2002) Identification of novel Helicobacter spp. from a beluga whale. Appl Environ Microbiol 68:2040–2043
14. Taylor NS, Xu S, Nambiar P et al (2007) Enterohepatic Helicobacter species are prevalent in mice from commercial and academic institutions in Asia, Europe, and North America. J Clin Microbiol 45:2166–2172
15. Fox JG, Edrise BM, Cabot EB et al (1986) Campylobacter-like organisms isolated from gastric mucosa of ferrets. Am J Vet Res 47:236–239
16. Gottfried MR, Washington K, Harrell LJ (1990) Helicobacter pylori-like microorganisms and chronic active gastritis in ferrets. Am J Gastroenterol 85:813–818
17. Croinin TO, McCormack A, van Vliet AH et al (2007) Random mutagenesis to identify novel Helicobacter mustelae virulence factors. FEMS Immunol Med Microbiol 50:257–263
18. Fox JG, Cabot EB, Taylor NS et al (1988) Gastric colonization by Campylobacter pylori subsp. mustelae in ferrets. Infect Immun 56:2994–2996
19. Fox JG, Chilvers T, Goodwin C (1989) Campylobacter mustelae, a new species resulting from the elevation of Campylobacter pylori subsp. mustelae to species status. Int J Syst Bacteriol 39:301–303
20. Perkins SE, Fox JG, Walsh JH (1996) Helicobacter mustelae-associated hypergastrinemia in ferrets (Mustela putorius furo). Am J Vet Res 57:147–150
21. Fox JG, Dangler CA, Sager W et al (1997) Helicobacter mustelae-associated gastric adenocarcinoma in ferrets (Mustela putorius furo). Vet Pathol 34:225–229
22. Yu J, Russell RM, Salomon RN et al (1995) Effect of Helicobacter mustelae infection on ferret gastric epithelial cell proliferation. Carcinogenesis 16:1927–1931
23. Correa P (1992) Human gastric carcinogenesis: a multistep and multifactorial process–first American cancer society award lecture on cancer epidemiology and prevention. Cancer Res 52:6735–6740
24. Fox JG (1995) Non-human reservoirs of Helicobacter pylori. Aliment Pharmacol Ther 9(Suppl 2):93–103

25. Lecoindre P, Chevallier M, Peyrol S et al (2000) Gastric helicobacters in cats. J Feline Med Surg 2:19–27
26. Newell DG, Lee A, Hawtin PR et al (1989) Antigenic conservation of the ureases of spiral- and helical-shaped bacteria colonising the stomachs of man and animals. FEMS Microbiol Lett 53:183–186
27. Handt LK, Fox JG, Dewhirst FE et al (1994) Helicobacter pylori isolated from the domestic cat: public health implications. Infect Immun 62:2367–2374
28. Otto G, Hazell SH, Fox JG et al (1994) Animal and public health implications of gastric colonization of cats by Helicobacter-like organisms. J Clin Microbiol 32:1043–1049
29. Handt LK, Fox JG, Stalis IH et al (1995) Characterization of feline Helicobacter pylori strains and associated gastritis in a colony of domestic cats. J Clin Microbiol 33:2280–2289
30. Fox JG, Batchelder M, Marini R et al (1995) Helicobacter pylori-induced gastritis in the domestic cat. Infect Immun 63:2674–2681
31. Perkins SE, Fox JG, Marini RP et al (1998) Experimental infection in cats with a cagA + human isolate of Helicobacter pylori. Helicobacter 3:225–235
32. Simpson KW, Strauss-Ayali D, Straubinger RK et al (2001) Helicobacter pylori infection in the cat: evaluation of gastric colonization, inflammation and function. Helicobacter 6:1–14
33. Straubinger RK, Greiter A, McDonough SP et al (2003) Quantitative evaluation of inflammatory and immune responses in the early stages of chronic Helicobacter pylori infection. Infect Immun 71:2693–2703
34. Esteves MI, Schrenzel MD, Marini RP et al (2000) Helicobacter pylori gastritis in cats with long-term natural infection as a model of human disease. Am J Pathol 156:709–721
35. Paster BJ, Lee A, Fox JG et al (1991) Phylogeny of Helicobacter felis sp. nov., Helicobacter mustelae, and related bacteria. Int J Syst Bacteriol 41:31–38
36. Simpson KW, Strauss-Ayali D, Scanziani E et al (2000) Helicobacter felis infection is associated with lymphoid follicular hyperplasia and mild gastritis but normal gastric secretory function in cats. Infect Immun 68:779–790
37. Bridgeford EC, Marini RP, Feng Y et al (2008) Gastric Helicobacter species as a cause of feline gastric lymphoma: a viable hypothesis. Vet Immunol Immunopathol 123:106–113
38. Munson L, Terio KA, Worley M et al (2005) Extrinsic factors significantly affect patterns of disease in free-ranging and captive cheetah (Acinonyx jubatus) populations. J Wildl Dis 41:542–548
39. Citino SB, Munson L (2005) Efficacy and long-term outcome of gastritis therapy in cheetahs (Acinonyx jubatus). J Zoo Wildl Med 36:401–416
40. Terio KA, Munson L, Marker L et al (2005) Comparison of Helicobacter spp. in Cheetahs (Acinonyx jubatus) with and without gastritis. J Clin Microbiol 43:229–234
41. Solnick JV, Chang K, Canfield DR et al (2003) Natural acquisition of Helicobacter pylori infection in newborn rhesus macaques. J Clin Microbiol 41:5511–5516
42. Solnick JV, Fong J, Hansen LM et al (2006) Acquisition of Helicobacter pylori infection in rhesus macaques is most consistent with oral-oral transmission. J Clin Microbiol 44:3799–3803
43. Solnick JV, Canfield DR, Yang S et al (1999) Rhesus monkey (Macaca mulatta) model of Helicobacter pylori: noninvasive detection and derivation of specific-pathogen-free monkeys. Lab Anim Sci 49:197–201
44. Solnick JV, Hansen LM, Canfield DR et al (2001) Determination of the infectious dose of Helicobacter pylori during primary and secondary infection in rhesus monkeys (Macaca mulatta). Infect Immun 69:6887–6892
45. Boonjakuakul JK, Canfield DR, Solnick JV (2005) Comparison of Helicobacter pylori virulence gene expression in vitro and in the Rhesus macaque. Infect Immun 73:4895–4904
46. Huff JL, Hansen LM, Solnick JV (2004) Gastric transcription profile of Helicobacter pylori infection in the rhesus macaque. Infect Immun 72:5216–5226
47. Hirayama F, Takagi S, Kusuhara H et al (1996) Induction of gastric ulcer and intestinal metaplasia in mongolian gerbils infected with *Helicobacter pylori*. J Gastroenterol 31:755–757
48. Peek RM Jr, Blaser MJ (2002) Helicobacter pylori and gastrointestinal tract adenocarcinomas. Nat Rev Cancer 2:28–37
49. Sawada Y, Yamamoto N, Sakagami T et al (1999) Comparison of pathologic changes in Helicobacter pylori-infected Mongolian gerbils and humans. J Gastroenterol 34(Suppl 11):55–60
50. Franco AT, Israel DA, Washington MK et al (2005) Activation of beta-catenin by carcinogenic *Helicobacter pylori*. Proc Natl Acad Sci USA 102:10646–10651
51. Lee A (2000) Animal models of gastroduodenal ulcer disease. Baillieres Best Pract Res Clin Gastroenterol 14:75–96

52. Bergin IL, Sheppard BJ, Fox JG (2003) Helicobacter pylori infection and high dietary salt independently induce atrophic gastritis and intestinal metaplasia in commercially available outbred Mongolian gerbils. Dig Dis Sci 48:475–485
53. Lee A, Fox JG, Otto G et al (1990) A small animal model of human Helicobacter pylori active chronic gastritis. Gastroenterology 99:1315–1323
54. De Bock M, D'Herde K, Duchateau L et al (2006) The effect of Helicobacter felis and Helicobacter bizzozeronii on the gastric mucosa in Mongolian gerbils: a sequential pathological study. J Comp Pathol 135:226–236
55. Gasbarrini A, Carloni E, Gasbarrini G et al (2003) Helicobacter pylori and extragastric diseases – other helicobacters. Helicobacter 8(Suppl 1):68–76
56. Fox JG, Sheppard BJ, Dangler CA et al (2002) Germ-line p53-targeted disruption inhibits helicobacter-induced premalignant lesions and invasive gastric carcinoma through down- regulation of Th1 proinflammatory responses. Cancer Res 62:696–702
57. Fox JG, Lee A: Gastric helicobacter infection in animals: natural and experimental infections. Edited by Goodwin C, Worsley B. Boca Raton, CRC Press, 1993, p. pp. 407–430
58. Mohammadi M, Redline R, Nedrud J et al (1996) Role of the host in pathogenesis of Helicobacter-associated gastritis: H. felis infection of inbred and congenic mouse strains. Infect Immun 64:238–245
59. Sakagami T, Shimoyama T, O'Rourke J et al (1994) Back to the host: severity of inlfammation induced by Helicobacter felis in different strains of mice (abstract). Am J Gastroenterol 89:1345
60. Fox JG (1998) Review article: Helicobacter species and in vivo models of gastrointestinal cancer. Aliment Pharmacol Ther 12(Suppl 1):37–60
61. Lee A, O'Rourke J, De Ungria MC et al (1997) A standardized mouse model of Helicobacter pylori infection: introducing the Sydney strain. Gastroenterology 112:1386–1397
62. Fox JG, Wang TC, Rogers AB et al (2003) Host and microbial constituents influence Helicobacter pylori-induced cancer in a murine model of hypergastrinemia. Gastroenterology 124:1879–1890
63. Rogers AB, Taylor NS, Whary MT et al (2005) Helicobacter pylori but not high salt induces gastric intraepithelial neoplasia in B6129 mice. Cancer Res 65:10709–10715
64. Touati E, Michel V, Thiberge JM et al (2003) Chronic Helicobacter pylori infections induce gastric mutations in mice. Gastroenterology 124:1408–1419
65. Fox JG, Rogers AB, Ihrig M et al (2003) Helicobacter pylori-associated gastric cancer in INS-GAS mice is gender specific. Cancer Res 63:942–950
66. Wang TC, Dangler CA, Chen D et al (2000) Synergistic interaction between hypergastrinemia and Helicobacter infection in a mouse model of gastric cancer. Gastroenterology 118:36–47
67. Stenstrom B, Zhao CM, Rogers AB et al (2007) Swedish moist snuff accelerates gastric cancer development in Helicobacter pylori-infected wild-type and gastrin transgenic mice. Carcinogenesis 28(9):2041–2046
68. Ohtani M, Garcia A, Rogers AB et al (2007) Protective role of 17{beta}-estradiol against the development of *Helicobacter pylori*-induced gastric cancer in INS-GAS mice. Carcinogenesis 28:2597–2604
69. Takaishi S, Tu S, Dubeykovskaya ZA et al (2009) Gastrin is an essential cofactor for helicobacter-associated gastric corpus carcinogenesis in C57BL/6 mice. Am J Pathol 175: 365–375
70. Kuzushita N, Rogers AB, Monti NA et al (2005) p27kip1 deficiency confers susceptibility to gastric carcinogenesis in Helicobacter pylori-infected mice. Gastroenterology 129:1544–1556
71. Goldenring JR, Nomura S (2006) Differentiation of the gastric mucosa III. Animal models of oxyntic atrophy and metaplasia. Am J Physiol Gastrointest Liver Physiol 291:G999–G1004
72. Kang W, Rathinavelu S, Samuelson LC et al (2005) Interferon gamma induction of gastric mucous neck cell hypertrophy. Lab Invest 85:702–715
73. Nomura S, Baxter T, Yamaguchi H et al (2004) Spasmolytic polypeptide expressing metaplasia to preneoplasia in *H. felis*-infected mice. Gastroenterology 127:582–594
74. Rogers AB, Houghton J (2009) Helicobacter-based mouse models of digestive system carcinogenesis. Methods Mol Biol 511:267–295
75. Fox JG, Rogers AB, Whary MT et al (2007) Accelerated progression of gastritis to dysplasia in the pyloric antrum of TFF2$^{-/-}$ C57BL6 × Sv129 *Helicobacter pylori*-infected mice. Am J Pathol 171:1520–1528
76. Kurt-Jones EA, Cao L, Sandor F et al (2007) Trefoil family factor 2 is expressed in murine gastric and immune cells and controls both

gastrointestinal inflammation and systemic immune responses. Infect Immun 75:471–480
77. Halldorsdottir AM, Sigurdardottrir M, Jonasson JG et al (2003) Spasmolytic polypeptide-expressing metaplasia (SPEM) associated with gastric cancer in Iceland. Dig Dis Sci 48:431–441
78. Harris CC (1996) p53 tumor suppressor gene: from the basic research laboratory to the clinic–an abridged historical perspective. Carcinogenesis 17:1187–1198
79. Fox JG, Li X, Cahill RJ et al (1996) Hypertrophic gastropathy in Helicobacter felis-infected wild-type C57BL/6 mice and p53 hemizygous transgenic mice. Gastroenterology 110:155–166
80. Donehower LA, Harvey M, Slagle BL et al (1992) Mice deficient for p53 are developmentally normal but susceptible to spontaneous tumours. Nature 356:215–221
81. Fox JG, Li X, Yan L et al (1996) Chronic proliferative hepatitis in A/JCr mice associated with persistent Helicobacter hepaticus infection: a model of helicobacter- induced carcinogenesis. Infect Immun 64:1548–1558
82. Harvey M, McArthur MJ, Montgomery CA Jr et al (1993) Spontaneous and carcinogen-induced tumorigenesis in p53-deficient mice. Nat Genet 5:225–229
83. Tomita H, Yamada Y, Oyama T et al (2007) Development of gastric tumors in Apc(Min/+) mice by the activation of the beta-catenin/Tcf signaling pathway. Cancer Res 67: 4079–4087
84. Moser AR, Luongo C, Gould KA et al (1995) ApcMin: a mouse model for intestinal and mammary tumorigenesis. Eur J Cancer 31A:1061–1064
85. Yang K, Edelmann W, Fan K et al (1997) A mouse model of human familial adenomatous polyposis. J Exp Zool 277:245–254
86. Fox JG, Dangler CA, Whary MT et al (1997) Mice carrying a truncated Apc gene have diminished gastric epithelial proliferation, gastric inflammation, and humoral immunity in response to Helicobacter felis infection. Cancer Res 57:3972–3978
87. De Sablet T, Piazuelo MB, Shaffer CL et al (2010) Phylogeographic origin of Helicobacter pylori is a determinant of gastric cancer risk. Gut 60(9):1189–1195
88. Chandanos E, Lagergren J (2008) Oestrogen and the enigmatic male predominance of gastric cancer. Eur J Cancer 44:2397–2403
89. Ghoshal UC, Chaturvedi R, Correa P (2010) The enigma of Helicobacter pylori infection and gastric cancer. Indian J Gastroenterol 29:95–100
90. Rogers AB, Fox JG (2004) Inflammation and Cancer. I. Rodent models of infectious gastrointestinal and liver cancer. Am J Physiol Gastrointest Liver Physiol 286:G361–G366
91. Rao VP, Poutahidis T, Ge Z et al (2006) Proinflammatory CD4+CD45RBhi lymphocytes promote mammary and intestinal carcinogenesis in ApcMin/+ mice. Cancer Res 66:57–61
92. Fox JG, Feng Y, Theve EJ et al (2010) Gut microbes define liver cancer risk in mice exposed to chemical and viral transgenic hepatocarcinogens. Gut 59:88–97
93. Brenner H, Rothenbacher D, Arndt V (2009) Epidemiology of stomach cancer. Methods Mol Biol 472:467–477
94. Whary MT, Sundina N, Bravo LE et al (2005) Intestinal helminthiasis in Colombian children promotes a Th2 response to *Helicobacter pylori:* possible implications for gastric carcinogenesis. Cancer Epidemiol Biomarkers Prev 14:1464–1469
95. Abou Holw SA, Anwar MM, Bassiouni RB et al (2008) Impact of coinfection with Schistosoma mansoni on Helicobacter pylori induced disease. J Egypt Soc Parasitol 38:73–84
96. Holcombe C (1992) Helicobacter pylori: the African enigma. Gut 33:429–431
97. Agha A, Graham DY (2005) Evidence-based examination of the African enigma in relation to Helicobacter pylori infection. Scand J Gastroenterol 40:523–529
98. Henderson KS (2008) Murine norovirus, a recently discovered and highly prevalent viral agent of mice. Lab Anim (NY) 37:314–320

Chapter 22

Histologic Scoring of Gastritis and Gastric Cancer in Mouse Models

Arlin B. Rogers

Abstract

Histopathology is a defining endpoint in mouse models of experimental gastritis and gastric adenocarcinoma. Presented here is an overview of the histology of gastritis and gastric cancer in mice experimentally infected with *Helicobacter pylori* or *H. felis*. A modular histopathologic scoring scheme is provided that incorporates relevant disease-associated changes. Whereas the guide uses *Helicobacter* infection as the prototype challenge, features may be applied to chemical and genetically engineered mouse models of stomach cancer as well. Specific criteria included in the combined gastric histologic activity index (HAI) include inflammation, epithelial defects, oxyntic atrophy, hyperplasia, pseudopyloric metaplasia, and dysplasia or neoplasia. Representative photomicrographs accompany descriptions for each lesion grade. Differentiation of genuine tumor invasion from pseudoinvasion is highlighted. A brief comparison of normal rodent versus human stomach anatomy and physiology is accompanied by an introduction to mouse-specific lesions including mucous metaplasia and eosinophilic droplets (hyalinosis). In conjunction with qualified pathology support, this guide is intended to assist research scientists, postdoctoral fellows, graduate students, and medical professionals from affiliated disciplines in the interpretation and histologic grading of chronic gastritis and gastric carcinoma in mouse models.

Key words: Comparative histology, Comparative anatomy, Gastric neoplasms, *Helicobacter pylori*, Mice

1. Introduction

A requirement of faithful animal models is the ability to reproduce the pathologic features that define the human disease. Fortunately, mice infected experimentally with *H. felis* and some *H. pylori* strains recapitulate the well characterized human "precancerous cascade" associated with *H. pylori* infection of humans as first described by Pelayo Correa [1]. The canonical histologic progression in humans is (1) chronic active nonatrophic gastritis, (2) multifocal atrophic

JeanMarie Houghton (ed.), *Helicobacter Species: Methods and Protocols*, Methods in Molecular Biology, vol. 921,
DOI 10.1007/978-1-62703-005-2_22, © Springer Science+Business Media, LLC 2012

gastritis, (3) intestinal metaplasia (complete followed by incomplete), (4) dysplasia, and (5) invasive carcinoma. Luckily, few of the nearly 50% of humans infected with *H. pylori* progress through all five stages (2). In mice, disease likewise is modulated by host genetics, microbial virulence, and the environment. Mice that mount strong Th1-type immune responses such as C57BL/6 develop chronic gastritis ± cancer following pathogenic *Helicobacter* infection, whereas those with a strong Th2 bias such as BALB/c may display a lymphofollicular proliferative disease resembling human MALT lymphoma (3). Still others exhibit immune indifference to gastric *Helicobacter* spp., maintaining high levels of bacterial colonization with little overt disease (4). As covered elsewhere in this book, *H. felis* is a better mouse-adapted pathogen than *H. pylori*. For example, wild-type C57BL/6 mice infected with *H. felis* may progress from chronic gastritis to adenocarcinoma, whereas comparable infection with *H. pylori* produces only non-malignant epithelial dysplasia (5, 6). This chapter details a histologic scoring scheme for the semi-quantitative assessment of chronic gastritis and dysplasia/neoplasia in mouse models. Although based on *Helicobacter*, this modular scoring scheme may be adapted to other experimental systems that invoke gastritis and/or stomach neoplasms. Unique anatomic features and species-specific lesions of the mouse stomach are introduced, and their potential implications in experimental disease discussed. The intention of this overview is to assist nonpathologists in the histologic evaluation of experimental gastritis and gastric carcinogenesis in mice, and to introduce a common language to help scientists from different laboratories meaningfully compare results.

1.1. Precancerous Cascade in Humans

In humans, *H. pylori* infection produces a low-grade, usually subclinical inflammatory response referred to as chronic active nonatrophic gastritis (7). Fortunately, in many the disease does not progress past this stage. However, a subset of infected individuals will develop additional lesions that represent a potential risk for cancer development. The next histologic step in the human precancerous cascade, multifocal atrophic gastritis, is defined by segmental loss of mucosal glands in the antrum. Because of its scattered distribution, endoscopists routinely collect multiple biopsies to avoid missing an important lesion. Multifocal gastric atrophy often presents at the junction of the corpus and antrum along the lesser curvature at the incisura angularis (8). Glandular loss by necrosis and apoptosis is offset by compensatory proliferation. This can result in thickening of the mucosa that belies the "atrophic gastritis" moniker (3). As more glands are lost, the potential space is filled by fibrous connective tissue and infiltrating leukocytes, disrupting normal epithelial-stromal crosstalk (8). Scrambled intercellular queues contribute to the next stage of disease, intestinal metaplasia, in which gastric epithelial cells take on a morphologic and molecular phenotype consistent with small bowel (complete) or large bowel (incomplete) mucosa. These changes are associated with expression of intestinal-type acidic mucins and the enterocyte-specific nuclear protein Cdx2 (9). As in the case of Barrett's esophagus, inappropriate appearance of intestinal-type cells in the stomach is a hallmark of preneoplasia. A third type of epithelial metaplasia, pseudopyloric metaplasia, is defined by antralization of glands in the oxyntic mucosa (fundus and corpus). In humans, this process also is called spasmolytic polypeptide-enhancing metaplasia (SPEM), based on associations with the antral cell marker spasmolytic polypeptide (trefoil-factor 2; TFF2) (10). As discussed below, pseudopyloric metaplasia rather than intestinal metaplasia is the variety most often observed in mouse models (11). The final stages of the cascade: dysplasia, in situ carcinoma, and invasive adenocarcinoma (ACA), are equivalent to those found in other epithelial cancers.

1.2. Precancerous Cascade in Mice

The same histologic stages from chronic inflammation to cancer largely are recapitulated in mice (Fig. 1a) (5). As in humans, murine chronic nonatrophic gastritis progresses to atrophic gastritis along the lesser curvature in a multifocal pattern. For this reason it is imperative to collect multiple gastric specimens from mice for histology. *Helicobacter*-associated lesions in mice may appear at either the proximal or distal margin of the corpus (i.e., the cardia or proximal antrum). Unlike humans, disease usually progress centrally such that the corpus rather than antrum is the primary target of mucosal alterations (3). As such, oxyntic rather than antral gland atrophy characterizes *Helicobacter* infection in the mouse. Nevertheless, some degree of atrophy of both oxyntic and antral glands will occur with gastric *Helicobacter* infection in both species. Atrophy of the oxyntic mucosa is significant because it may lead to loss of intrinsic factor from chief cells and achlorhydria from loss of parietal cells, permitting colonization of acid-sensitive microbial opportunists that may accelerate proinflammatory tumorigenesis (5). As described above, pseudopyloric metaplasia is much more common than intestinal metaplasia in mice. Pseudopyloric metaplasia must be distinguished histologically from mucous metaplasia in mice (discussed later), a morphologically distinct phenomenon characterized by replacement of parietal cells with $TFF2^{+}$ foamy cells resembling duodenal Brunner's glands (Fig. 1b) (11). The final stages of dysplasia, in situ carcinoma, and invasive ACA are similar between species. A consensus panel of pathologists and basic scientists has proposed the term gastrointestinal intraepithelial neoplasia (GIN) for dysplastic lesions of the murine mucosa, with subclassification into low-grade and high-grade based on degree of atypia (12). This panel considers GIN as synonymous with carcinoma in situ and host of other pathologic descriptors. Whereas the system was developed primarily for the

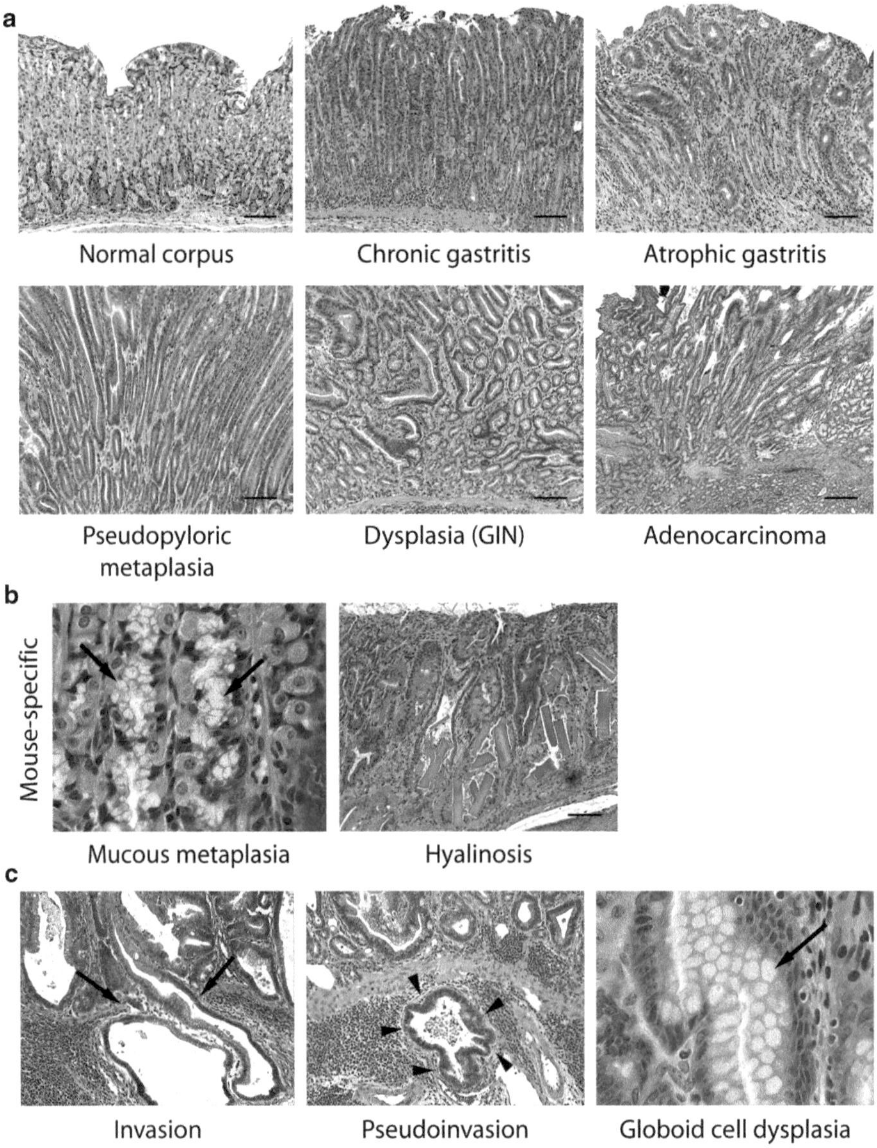

Fig. 1. Histopathology of gastric *Helicobacter* infection in the mouse. (**a**) Precancerous cascade in the mouse largely mirrors the sequence described by Correa in humans (8) including chronic gastritis, atrophic gastritis, pseudopyloric metaplasia (in humans intestinal metaplasia is more common), dysplasia (also known as gastrointestinal intraepithelial neoplasia or GIN), and invasive adenocarcinoma. (**b**) Mouse-specific gastric lesions that may occur either in the presence or absence of *Helicobacter* infection including mucous metaplasia and hyalinosis (*left* and *middle panels*). (**c**) Genuine tumor invasion is characterized by direct protrusion of dysplastic glands through the muscularis mucosae and into the submucosa (*arrows*; *left*). In contrast, pseudoinvasion is identified by the presence of orphan dysplastic glands in the submucosa beneath an intact muscularis mucosae (*middle*). Often these structures are lined by a thin rim of herniated muscularis mucosae (*arrowheads*). Globoid cell dysplasia (*right*) is characterized by clear expansion of atypical cell cytoplasm, frequent cell piling, and nuclear fading with cell extrusion. This change should not be confused with mucous metaplasia (panel (**b**)) or goblet cell metaplasia (very rare in mouse). Bar = 160 μm all panels except adenocarcinoma bar = 400 μm. Reproduced and revised with permission from Rogers and Houghton (5).

standardization of lower bowel tumor nomenclature, criteria are applicable to the stomach as well where the more focused name of gastric intraepithelial neoplasia has been applied (13). In the scoring scheme presented here, low-grade dysplasia is considered separately from high-grade dysplasia or GIN. Due to a limited life span, few *Helicobacter*-infected mice progress from GIN to unequivocal ACA. Reports of submucosal invasion often reflect pseudoinvasion, where herniated mucosal glands from an adjacent segment appear to be embedded in the submucosa of the tissue section under review (Fig. 1c; further discussed below). Distant metastasis is an exceedingly rare event in mouse models of gastric cancer.

1.3. Anatomic Considerations and Tissue Sampling of the Mouse Stomach

Whereas the mouse and human stomach have many features in common, there are important differences that must be recognized. The proximal third-to-half of the rodent stomach, including the anatomic equivalent of the human fundus, is lined by squamous rather than glandular epithelium. This appears grossly as a thin-walled chalk-white sac. Mice do not have "fundic glands," although the misnomer is ubiquitous in the scientific literature. In mice, functional gastric units containing parietal and chief cells are limited to the body of the stomach and are best referred to as corpus, oxyntic or zymogenic glands. The interface between the squamous forestomach and glandular stomach has been given different names including squamocolumnar junction and the forestomach/zymogenic junction (14). The band of glandular mucosa immediately adjacent to forestomach parallels in microscopic morphology the human cardia, but is composed of only 2–3 glandular units. Parietal and chief cells that define the beginning of the corpus appear soon thereafter. Unlike humans, oxyntic glands in the mouse ring the corpus discontinuously. Therefore, it is not uncommon in histologic sections to observe glands with an antral morphology extending the entire length of glandular stomach from squamocolumnar junction to pylorus (Fig. 2) (4). This must not be confused with oxyntic atrophy or pseudopyloric metaplasia (15). The region of interest to most investigators using mouse models of *Helicobacter* infection will be the corpus and adjoining segments, although tumors may arise preferentially at the pylorus in select models (11). When sectioning the stomach, an incision is made along the greater curvature from esophagus through proximal duodenum (Fig. 3) (5). Contents are rinsed with sterile saline and the stomach is laid flat on a card or cutting board. Two or three linear strips from the mid-section (lesser curvature) from squamocolumnar junction through proximal duodenum should be collected for histology. When possible, the sections should be fixed, submitted, and processed adhered to a flat surface such as an index card. The histology laboratory may then flip the sections on side during embedding to assure full mucosa-to-serosa representation. Gastric tissue strips may be submitted without support,

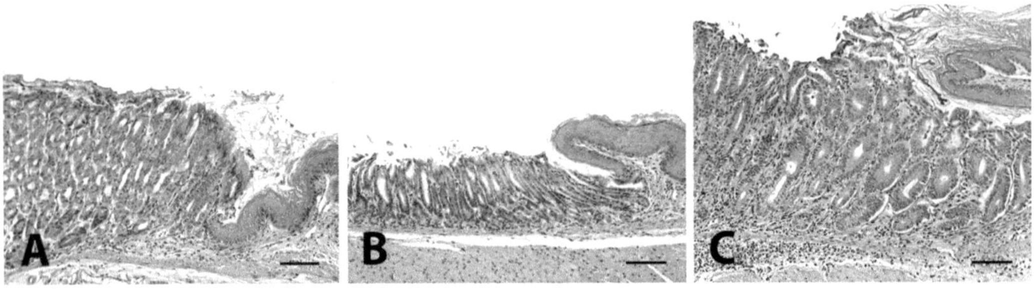

Fig. 2. Pseudopyloric metaplasia versus normal absence of oxyntic glands. (**a**) Normal stomach demonstrating squamous forestomach at right, squamocolumnar junction and glandular stomach at *left*. (**b**) Normal stomach with antral-type glands extending to the squamocolumnar junction highlighting the incompletely circumferential development of zymogenic glands in the corpus. (**c**) Pseudopyloric metaplasia characterized by loss of oxyntic mucosa and replacement by poorly differentiated glandular units with a more antral phenotype; note association with inflammation in this *H. pylori*-infected mouse. All panels bar = 160 μm. Reproduced with permission from Rogers and Fox (4).

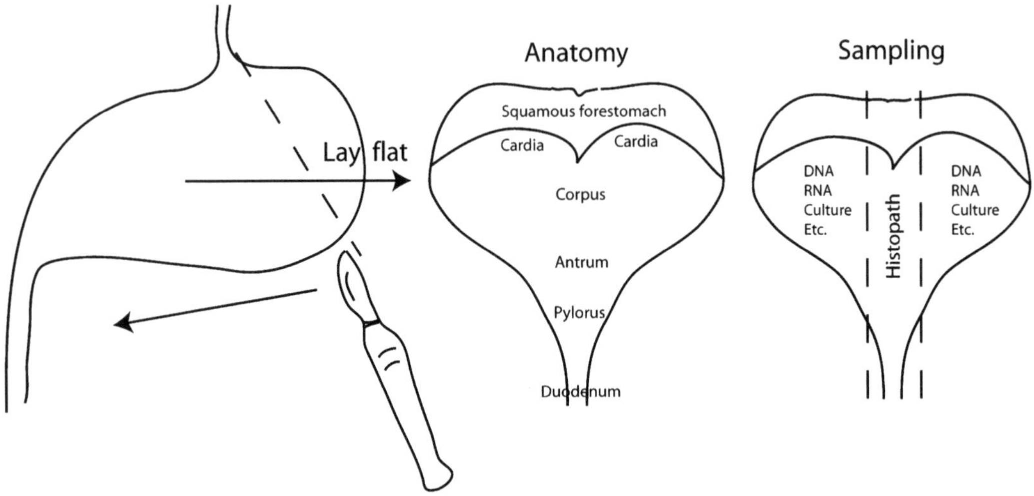

Fig. 3. Collection of stomach sections for histology and molecular analysis. An incision is made along the greater curvature, contents rinsed, and the stomach laid flat on a card. Multiple (2–3) linear strips from middle section representing lesser curvature should be collected from the squamocolumnar junction through proximal duodenum. At the time of embedding the strip should be place on side to insure representation of all layers from mucosa through serosa in the final stained specimen. Reproduced with permission from Rogers and Houghton (5).

but twisting and oblique orientations may develop that hinder histologic interpretation.

2. Histologic Scoring of Mouse Gastritis and Cancer

Six criteria are included in the basic *Helicobacter*-associated mouse gastric histology activity index (HAI): (1) inflammation, (2) epithelial defects, (3) oxyntic atrophy, (4) hyperplasia, (5) pseudopyloric metaplasia, and (6) dysplasia ± neoplasia (Fig. 4). Not all of these features will be evident in every model. For statistical comparisons, some investigators consider each criterion separately whereas others combine all criteria into a single HAI. If the latter approach is used, only lesions relevant to the model in question should be included. Each of the six primary criteria is described below, with specific features used to assign a score on a 0–4 scale. Additional lesions that may or may not be included in the HAI are eosinophilic droplets (hyalinosis) and mucous metaplasia (discussed later). Care must be taken when assigning biologic significance to these changes, because they may occur spontaneously (5). Edema is sometimes evaluated as an additional lesion category. However, artifactual expansion of the submucosa during tissue fixation and processing can mirror edematous change. A careful comparison of control versus treated animals will help determine whether clear expansion of the submucosa ± mucosa represents genuine edema.

2.1. Inflammation

At the molecular level, a proinflammatory microenvironment may be invoked in the absence of overt morphologic changes (16). However, from a histopathologic standpoint, inflammation is defined by the visible accumulation of infiltrating leukocytes. Whereas both polymorphonuclear (PMN) and mononuclear cells contribute to human and murine chronic active gastritis, PMN represent a smaller fraction of the total inflammatory census in mice as compared with humans. In humans, intraepithelial neutrophils are an important prognostic feature and merit distinct consideration in the updated Sydney scoring system (17). In contrast, intraepithelial neutrophils are a less reliable disease marker in mice, although PMN may be widely scattered in the lamina propria. Lymphocytes accompanied by fewer macrophages dominate the inflammatory landscape in the mouse stomach. Large round cells with basophilic granules in hematoxylin and eosin (H&E)-stained sections also may be present. These are mast cells, which are more common in rodents than humans in mixed inflammation. Mast cells and oxyntic parietal cells autofluoresce under standard fluorescence microscopy conditions, and should not be confused with target-specific cell labeling when performing fluorescence immunohistochemistry or in situ hybridization of the stomach.

Scoring criteria. Patchy or multifocal small islands of inflammatory cells in the mucosa and/or submucosa receive a score of 1. As infiltrates begin to coalesce across multiple high-power (40× objective) microscopic fields, the score increases to 2. Expansile sheets and/or lymphoid follicles in the mucosa or submucosa merit a score of 3. Extension of florid inflammation into the muscularis propria ± adventitia (i.e., transmural inflammation) receives a score of 4. In infectious models of gastric carcinogenesis, epithelial changes tend to increase in tandem with inflammation severity. However, in chemical or transgenic models dysplasia and neoplasia, disease may advance in the absence of overt inflammation.

	1	2	3	4
Inflammation	Multifocal aggregates of mononuclear ± polymorphonuclear leukocytes (arrows)	Coalescing aggregates of inflammatory cells in submucosa ± mucosa (mononuclear predominant)	Organizing nodules of lymphocytes and other inflammatory cells in submucosa ± mucosa	Follicles and/or sheets of inflammatory cells extending into or through muscularis ± adventitia
Epith. defects	Tattered epithelial surface; occasional dilated glands with sloughed apoptotic cells	Distinctly attenuated surface epithelium; numerous ectatic glands and necrotic cells and debris	Inapparent surface epithelial lining; few if any recognizable gastric pits	Mucosal erosions with multifocal collapse and fibrosis; surface lined by fibrin and/or inflammatory cell debris
Oxyntic atrph.	Decreased chief cells at base of glands; parietal cell volume intact	Few or no chief cells; most parietal cells intact	No chief cells; significant loss of parietal cells	No chief cells; few if any parietal cells
Hyperplasia	Moderate elongation of gastric pits	Surface-type epithelium increased at least 2X normal length; some loss of oxyntic glands	3X or more elongation of surface-type epithelium with significant loss of zymogenic glands	4X or more elongation of surface-type or dysplastic epithelium; complete loss of functional mucosa
Pseudopyloric	Replacement of <50% oxyntic mucosa by surface-type or antralized gastric glands	Replacement of >50% oxyntic mucosa by surface-type or antralized gastric glands	Near complete replacement of oxyntic mucosa by antralized glands	Total replacement of zymogenic glands by antralized mucosa
Dysplasia	Multifocal dysplastic glands characterized by elongation, altered shapes, back-to-back forms, etc.	Coalescing dysplasia with glandular ectasia, branching, infolding, cell piling up, globoid cells, etc.	Gastric intraepithelial neoplasia (GIN); loss of normal architecture and columnar orientation; atypical hyperchromatic cells	Invasive adenocarcinoma extending into submucosa, vessels, lymphatics, and/or deeper layers

Fig. 4. Histologic scoring scheme for murine gastritis and cancer. Descriptions and photomicrographs of representative lesions on a 1–4 scale for each of the five primary criteria that factor into the HAI are shown. Categories include inflammation, epithelial defects, oxyntic atrophy, hyperplasia, and dysplasia/neoplasia. All images 20× objective.

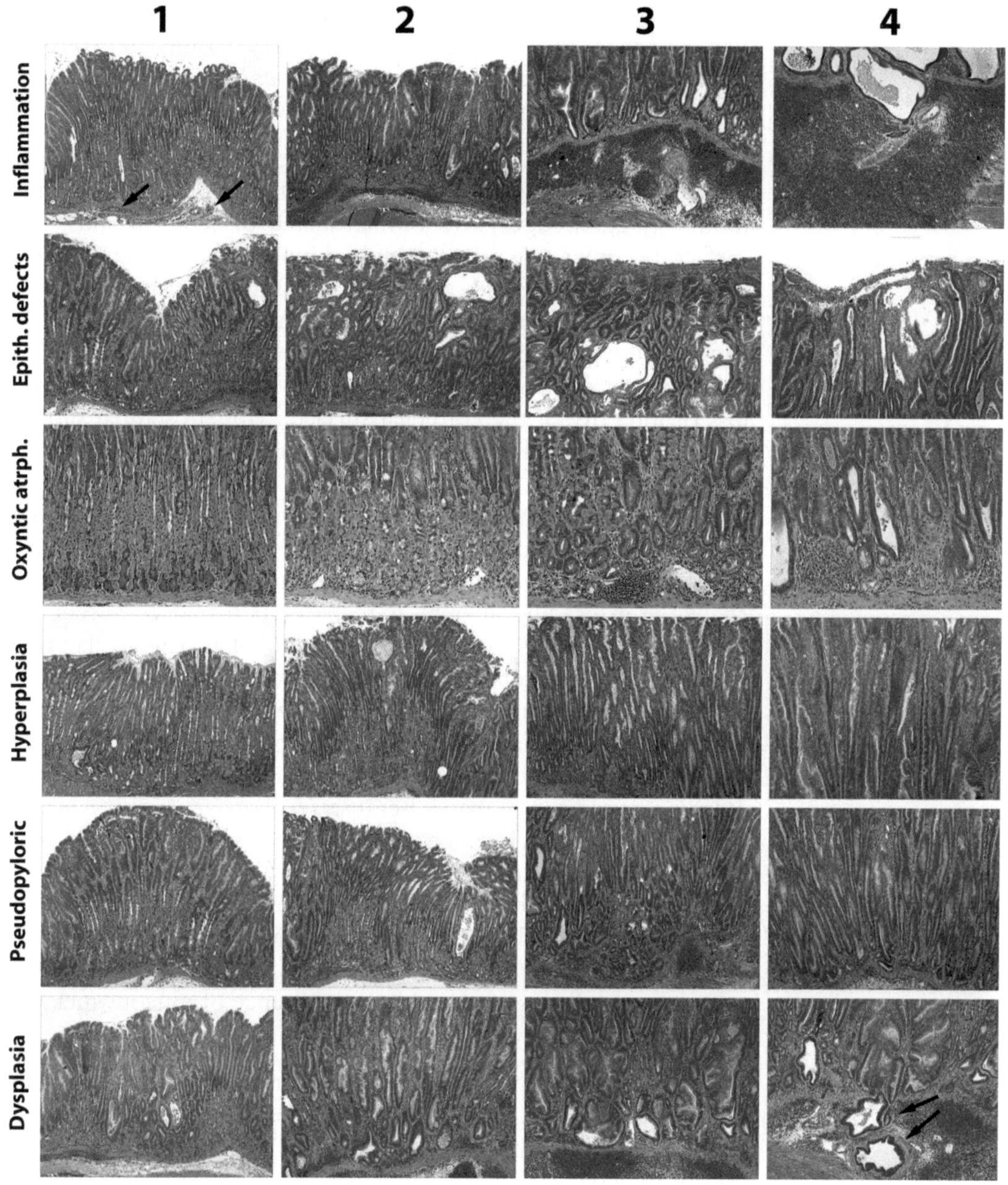

Fig. 4. (continued)

2.2. Epithelial Defects

Humans with clinically overt *H. pylori* infection usually develop one of two disease presentations: peptic ulcer disease or atrophic gastritis ± metaplasia/neoplasia (2). These phenotypes are rather exclusive, and rarely develop in the same individual. Similarly in animals, models associated with the inflammation–metaplasia–neoplasia sequence do not exhibit peptic ulcer disease. Therefore, the highest grade for epithelial defects in the present scoring system is assigned to surface erosions and not full-thickness ulceration, although there may also be subsurface gland atrophy and collapse. Overall, gastric mucosal defects in *Helicobacter*-infected mice are less severe in mice than those found in other animal models such as Mongolian gerbils and ferrets (4).

Scoring criteria. A score of 1 is assigned to surface epithelial "tattering" (jagged edges) ± occasional dilated or ectatic glands. The surface becomes more attenuated (thinned) in a grade 2 lesion, and dilated ± cystic glands are common. Only a very thin epithelial surface lining remains with a grade 3, gland ectasia may be widespread, and often there is some atrophy of subjacent mucosal glands with compensatory fibroplasia. Frank erosions leading to a loss of integrity of the epithelial surface, with substantial subsurface gland atrophy and fibrosis earn a score of 4.

2.3. Oxyntic Atrophy

The oxyntic mucosa, defined by the presence of parietal and chief cells (i.e., "fundic glands"), is limited to the gastric corpus in the mouse, and even there may be discontinuous. Therefore, the first challenge in grading atrophy is to be certain that an absence of parietal and chief cells in a given section of proximal glandular stomach is not normal. In the absence of accompanying lesions such as inflammation or dysplasia, an antralized mucosa adjacent to the squamocolumnar junction should be disregarded (Fig. 2). Additionally, oxyntic atrophy of mice should not be confused with atrophic gastritis in humans. Because *Helicobacter*-induced lesions in the mouse typically emerge at the proximal and/or distal margins of the corpus and progress centrally, antral atrophy is rarely a significant part of the murine disease profile. Oxyntic gland atrophy, in contrast, is always apparent in mouse models of progressive gastritis and cancer. This is due in part to the fact that loss of chief and parietal cells is an inevitable consequence of mucous metaplasia and/or pseudopyloric metaplasia.

Scoring criteria. Oxyntic atrophy in the mouse follows a predictable progression, with chief cell disappearance always preceding that of parietal cells. As such, a grade of 1 describes loss of about half of the chief cells, and a grade of 2 the near complete absence of chief cells but only a minimal loss of parietal cells. A score of 3 means that all chief cells are absent along with about half the expected mass of parietal cells, and a score of 4 signifies near total loss of both cell populations. Note that oxyntic atrophy is not correlated with the overall thickness of the affected mucosa. Indeed, atrophy of oxyntic and zymogenic glands often is accompanied by hyperplasia of replacement cells, resulting in increased overall mucosal thickness. Loss of antral glands with replacement fibrosis characterizing human atrophic gastritis does occur in the mouse (Fig. 1a), but is not a required step in the precancerous cascade.

2.4. Hyperplasia

Hyperplasia refers to elongation of gastric gland units due to increased numbers of surface (foveolar) and/or antral-type epithelial cells. Expansion of the oxyntic cell mass is *not* included in this category. Indeed, oxyntic hyperplasia is extremely uncommon in the murine stomach except for certain genetically engineered models with hypergastrinemia such as INS-GAS mice. In the early stages of *Helicobacter* infection, most epithelial hyperplasia will be represented by expansion of gastric pit cells in the upper third of the mucosa. As disease progresses, these cells may dedifferentiate into a generic columnar morphotype, and extend deeply into the oxyntic zone. By the time most of the parietal cells have been replaced, there often is concurrent mucous or pseudopyloric metaplasia. Whereas hyperplasia and dysplasia/neoplasia frequently increase in severity together, it is possible to have significant hyperplasia with retention of the normal columnar glandular orientation and cell populations. Conversely, dysplasia may proceed to in situ carcinoma even in areas of minimal hyperplasia.

Scoring criteria. Because hyperplasia is based on the expected thickness of the normal gastric mucosa, it is imperative to view a number of normal stomachs from control mice to establish a baseline. The thickness of the normal intact gastric mucosa will vary from model to model, and irregularities in tissue orientation may produce alternating areas with a thickened or thinned appearance. For this reason, manual cell counting and computer-assisted morphometrics provide little quantitative advantage over skilled observers. Only foveolar (surface) type cells, or antral-type cells in the case of pseudopyloric metaplasia, are factored into the hyperplasia grade. Because oxyntic atrophy frequently occurs concurrently, there may be significant hyperplasia in a mucosa of normal overall thickness. Compared to the expected length of gastric pits, a hyperplasia score of 1 implies an approximately 50% increase over the expected length. A score of 2 = 2× the expected length; 3 = 3×, and 4 = 4× the expected length. A score of up to 2 may occur in the absence of overall mucosal thickening if oxyntic glands have been lost in the process. A score of 3 or greater always implies overall thickening of the gastric mucosa.

2.5. Pseudopyloric Metaplasia

As discussed previously, pseudopyloric metaplasia is the replacement of the oxyntic mucosa by glands with an antral phenotype. Antralized cells are more columnar, and lack the internal granules and unique staining features characteristic of parietal and chief cells. Pseudopyloric metaplasia has the same preneoplastic significance in the mouse as intestinal metaplasia in humans. Whereas the human corpus and/or fundus also may acquire pseudopyloric metaplasia (known as SPEM, see above), the change is less reliable than intestinal metaplasia (10).

Scoring criteria. Pseudopyloric metaplasia is graded based on the amount of the gastric corpus replaced by antralized glands. When there is <25% replacement, the specimen receives a grade of 1. A grade 2 lesion exhibits 25–50% oxyntic mucosa replacement, grade 3 shows 50–75% replacement, and a grade 4 lesion demonstrates >75% replacement of the parietal and chief cell zone by antral-type glands. As discussed previously, care must be taken to insure that only preexisting oxyntic mucosa is considered in this paradigm, and not regions of the proximal gandular stomach that lacked parietal and chief cells to begin with (see Fig. 2). With few exceptions, pseudopyloric metaplasia will be associated with chronic inflammation, and a few parietal cells will be scattered amongst the metaplastic tissue as evidence of their prior occupancy.

2.6. Dysplasia/ Neoplasia

Atypical morphologic features of cells and glands that presage neoplastic transformation are similar between humans and mice. At the organizational level these include haphazard glandular arrangement, loss of vertical orientation, back-to-back associations without intervening stroma, branching and infolding, and cell piling up. At the cellular level dysplastic features include differences in overall cell and nuclear size (anisocytosis and anisokaryosis), hyperpleomorphism (highly variable cell shape and size), poorly defined cell junctions, loss of nuclear polarity, and hyperchromasia characterized by a tall columnar shape with increased nuclear–cytoplasmic (N–C) ratio (18). Epithelial cell crowding and hyperchromasia with increased N–C ratio factor heavily into the assessment of human gastric pinch biopsies, whereas in mice overall gland organization is a better indicator of dysplastic progression. Malignant transformation is assumed in cases with unequivocal invasion into the submucosa or beyond. However, true invasion must be differentiated from herniation or pseudoinvasion created by the entrapment of orphan mucosal glands from an adjacent segment of stomach that appear as free-floating "invasive glands" beneath an intact muscularis mucosae. Careful observation at high magnification usually reveals a thin lining of muscularis mucosae bounding such glandular islands. Unless direct continuance through the muscularis mucosae can be confirmed, histologic identification of isolated, well-circumscribed gastric glands in the submucosa should not be construed as proof of invasion.

Scoring criteria. Early dysplastic changes in the stomach that merit a dysplasia score of 1 resemble aberrant crypt foci in the colon. Features include distortion of normal columnar orientation, increased diameter, asymmetrical cell piling, and back-to-back forms. As these lesions coalesce and advance to grade 2, there is glandular infolding, branching, and more advanced cellular atypia such as increased N–C ratio. When normal parallel columnar orientation of glands is completely lost and there is marked glandular and cellular distortion with haphazard arrangements, the lesion qualifies as GIN or carcinoma in situ. Frequently round-to-ovoid cells with cytoplasmic clearing will pile up in some glands during this process. This is globoid cell dysplasia (Fig. 2c), not to be confused with signet ring cells or goblet cell metaplasia. Invasion into but not through the muscularis mucosae may be called intramucosal carcinoma, although this diagnosis is not universally recognized. A grade 4 lesion implies unequivocal invasion of highly dysplastic glands into the submucosa or beyond. To avoid misclassification of pseudoinvasion, documentation of direct invasion through the muscularis mucosae is preferred over the identification of isolated dysplastic glands in the submucosa (Fig. 2c). Vascular or lymphatic invasion is noteworthy, but caution must be exercised in this interpretation, as RBC within the capsule of dysplastic glands may give the appearance of a tumor embolus, and noninvasive glands may impinge on thin-walled vessels or lymphatics.

3. Mouse-Specific Gastric Lesions

In the mouse there are two species-specific lesions not described in humans: mucous metaplasia and eosinophilic droplets or hyalinosis (Fig. 1b). These may occur together or separately, and may appear spontaneously or in association with gastric *Helicobacter* infection. Different strains of mice exhibit different propensities to the development of these lesions, with 129S and C57BL/6 strains being among the susceptible. Recognition of these lesions is essential in the interpretation of disease induced by gastric *Helicobacter* infection in mouse models.

3.1. Mucous Metaplasia

Mucous metaplasia describes the replacement of oxyntic parietal and chief cells in the corpus with an expanded population of round foamy cells resembling those of duodenal Brunner's glands (13). When special mucin stains such as pH 2.5 Alcian blue/periodic acid-Schiff (AB/PAS) are applied, this foamy cell population expresses a mixture of normal neutral gastric-type mucins (red) in increased abundance, and acidic intestinal-type mucins (blue) normally found only in antral glands and/or the intestine. Because these large foamy cells replace resident chief and parietal cells, mucous metaplasia is always associated with oxyntic atrophy. As mentioned previously, this cell population expresses all of the molecular markers of SPEM even though it is not of comparable biologic significance (11). If the change is determined to be a component of the experimental disease phenotype, scores may be assigned based on percent target mucosa affected such that <25% corpus involvement receives a grade of 1, 25–50% a grade of 2, 50–75% a grade of 3, and >75% a grade of 4.

3.2. Eosinophilic Droplets (Hyalinosis)

Bright red round or crystalline structures in the murine gastric surface epithelial layer are known as eosinophilic droplets or hyalinosis. This substance is composed of a chitinase-like protein (Ym2) encoded by the gene chitinase 3-like 4 (19). A few droplets may be found normally in the cardia, but in extreme cases hyalin droplets and crystals may extend the full length of the corpus, and deeply into the glands. The function of this material is unclear, but it appears to represent a nonspecific response to epithelial injury (19). If scored, the percentage of target mucosa affected is the primary criterion, similar to mucous metaplasia. Eosinophilic droplets also may be found within intrahepatic bile ducts and the gallbladder. In the lung, hyaline material of a different chemical composition is associated with crystal pneumonitis or eosinophilic macrophage pneumonia (19, 20). Like mucous metaplasia, hyalinosis may develop either spontaneously or in association with *Helicobacter* infection, especially in mice on a 129S or C57BL/6 strain background. To further confound the picture, in cases where mucous metaplasia and hyalinosis occur concurrently, there may be a prominent eosinophil-predominant inflammatory infiltrate with epithelial erosions, hyperplasia and dysplasia. This can significantly complicate murine studies of *Helicobacter*-induced gastritis and carcinogenesis. To correctly interpret the overall disease presentation, it is always best for an experienced comparative pathologist to review *Helicobacter*-associated gastric lesions in mouse models (21).

4. Summary

Correct interpretation of morphologic outcomes is critical in experimental mouse models of gastric inflammation and carcinogenesis. Uniformity in scoring criteria is needed to improve comparison of results between different laboratories. This guide is intended to assist scientific investigators and medical professionals in understanding and objectively scoring histologic disease progression in mouse models. Lessons learned from these models will advance our understanding of mechanisms underlying *H. pylori* gastritis and cancer in humans, and accelerate the development of new biomarkers and therapies to prevent and treat this deadly disease.

References

1. Correa P (1985) Mechanisms of gastric carcinogenesis. In: Joossens JV, Hill MJ, Geboers J (eds) Diet and human carcinogenesis. Elsevier, Amsterdam, The Netherlands), pp 109–115
2. Wroblewski LE, Peek RM Jr, Wilson KT (2010) Helicobacter pylori and gastric cancer: factors that modulate disease risk. Clin Microbiol Rev 23:713–739
3. Rogers AB, Fox JG (2004) Inflammation and Cancer I. Rodent models of infectious gastrointestinal and liver cancer. Am J Physiol Gastrointest Liver Physiol 286:G361–G366
4. Rogers AB, Fox JG (2009) Animal models of gastric carcinoma. In: Giraud AS, Fox JG, Wang TC (eds) The biology of gastric cancers. Springer, New York, pp 323–360
5. Rogers AB, Houghton J (2009) Helicobacter-based mouse models of digestive system carcinogenesis. Methods Mol Biol 511:267–295
6. Houghton J, Stoicov C, Nomura S et al (2004) Gastric cancer originating from bone marrow-derived cells. Science 306:1568–1571
7. Correa P (1995) Helicobacter pylori and gastric carcinogenesis. Am J Surg Pathol 19(Suppl 1):S37–S43
8. Correa P, Houghton J (2007) Carcinogenesis of Helicobacter pylori. Gastroenterology 133:659–672
9. Barros R, Camilo V, Pereira B et al (2010) Pathophysiology of intestinal metaplasia of the stomach: emphasis on CDX2 regulation. Biochem Soc Trans 38:358–363
10. Goldenring JR, Nomura S (2006) Differentiation of the gastric mucosa III Animal models of oxyntic atrophy and metaplasia. Am J Physiol Gastrointest Liver Physiol 291:G999–G1004
11. Fox JG, Rogers AB, Whary MT et al (2007) Accelerated Progression of Gastritis to Dysplasia in the Pyloric Antrum of TFF2$^{-/-}$ C57BL6 x Sv129 *Helicobacter pylori*-Infected Mice. Am J Pathol 171:1520–1528
12. Boivin GP, Washington K, Yang K et al (2003) Pathology of mouse models of intestinal cancer: consensus report and recommendations. Gastroenterology 124:762–777
13. Rogers AB, Taylor NS, Whary MT et al (2005) Helicobacter pylori but not high salt induces gastric intraepithelial neoplasia in B6129 mice. Cancer Res 65:10709–10715
14. Syder AJ, Oh JD, Guruge JL et al (2003) The impact of parietal cells on Helicobacter pylori tropism and host pathology: an analysis using gnotobiotic normal and transgenic mice. Proc Natl Acad Sci USA 100: 3467–3472
15. Kang W, Rathinavelu S, Samuelson LC et al (2005) Interferon gamma induction of gastric mucous neck cell hypertrophy. Lab Invest 85:702–715
16. Fox JG, Feng Y, Theve EJ et al (2010) Gut microbes define liver cancer risk in mice exposed to chemical and viral transgenic hepatocarcinogens. Gut 59:88–97
17. Stolte M, Meining A (2001) The updated Sydney system: classification and grading of gastritis as the basis of diagnosis and treatment. Can J Gastroenterol 15:591–598
18. Fox JG, Rogers AB, Ihrig M et al (2003) Helicobacter pylori-associated gastric cancer in INS-GAS mice is gender specific. Cancer Res 63:942–950
19. Ward JM, Yoon M, Anver MR et al (2001) Hyalinosis and Ym1/Ym2 gene expression in the stomach and respiratory tract of 129 S4/SvJae and wild-type and CYP1A2-null B6, 129 mice. Am J Pathol 158:323–332
20. Haines DC, Chattopadhyay S, Ward JM (2001) Pathology of aging B6;129 mice. Toxicol Pathol 29:653–661
21. Cardiff RD, Ward JM, Barthold SW (2007) 'One medicine-one pathology': are veterinary and human pathology prepared? Lab Invest 88(1):18–26

Chapter 23

Innate Immune Responses to *Helicobacter pylori* Infection: An Overview

Milan K. Patel, Melanie I. Trombly, and Evelyn A. Kurt-Jones

Abstract

Innate immune receptors detect *Helicobacter pylori* infection and trigger downstream signaling events that result in the production of cytokines and interferon-β. This chapter gives an overview of the receptors and their roles in responding to *H. pylori* infection and details the downstream signaling events. The tools that have been developed to study the innate immune response to *H. pylori* are also discussed. Understanding the immune response to *H. pylori* is critical to develop better treatments for *H. pylori*-induced disease states including gastric malignancies and cancer.

Key words: Toll-like receptors, Nod-like receptors, AIM2-like receptors, Cytokines, ELISA, Luciferase, *Helicobacter pylori*

The innate immune system is the first line of defense against invading pathogens. The system consists of various receptors that activate proinflammatory pathways. Surface and endosomal receptors include Toll-like receptors (TLRs) that recognize various pathogen components including LPS, bacterial cell wall components, nucleic acids, and flagellin (1–3). In addition to surface and endosomal detection, there are many intracellular receptors that survey the cytoplasm for the presence of pathogenic particles. These include Nod-like receptors (NLRs) that recognize peptidoglycans and detect components released by damaged host cells. Other cytosolic receptors include the RIG-I-like (RLR) and AIM2-like (ALR) receptors that detect nucleic acids (1, 2). Recognition of pathogenic elements by TLRs and NLRs results in activation of NF-κB and MAPK pathways and induces the synthesis and secretion of inflammatory cytokines (1–8). Engagement of NLR and AIM2 receptors triggers inflammasome assembly and ASC/Caspase-1 activation leading to processing of pro-IL-1β and pro-IL-18 and the subsequent secretion of active IL-1β and IL-18 (2, 3, 8). The inflammatory response, however, is not always beneficial for the

JeanMarie Houghton (ed.), *Helicobacter Species: Methods and Protocols*, Methods in Molecular Biology, vol. 921,
DOI 10.1007/978-1-62703-005-2_23, © Springer Science+Business Media, LLC 2012

host. In the case of *Helicobacter pylori* infection, the resulting inflammation can lead to severe gastric immunopathology and cancer (1, 2, 9).

H. pylori initiates a robust immune response. TLR2 is the major innate receptor for the recognition of *H. pylori* infection and resulting inflammation. TLR2 has been shown to be the dominant receptor for intact *H. pylori* bacteria. TLR2 is also activated by *H. pylori* heat-shock protein 60 (3, 10, 11). TLR4 (LPS receptor) and TLR5 (flagellin receptor) have also been implicated in innate immune responses to *H. pylori* (3, 10, 11). In addition, the *cagA* secretion system of *H. pylori* delivers peptidoglycans into host cells activating intracellular receptors such as Nod1 (8). Activation of these innate receptors leads to activation of NF-κB, caspase, and interferon pathways that result in production of proinflammatory cytokines such as IL-1β, TNFα, IL-6, IL-8, MCP-1, and IFNβ (3, 8). These cytokines attract acute inflammatory mediators such as neutrophils as well as lymphocytes leading to activation of the adaptive immune response. Chronic infections with *H. pylori* lead to a persistent immune response by the host that contributes to gastric malignancies (1).

Understanding the immune responses initiated by *H. pylori* can give insights into the development of gastritis and subsequently *H. pylori*-induced disease states. Proinflammatory cytokines produced in response to infection are an excellent way to analyze the early innate immune response. Many commercial kits are available to measure production of inflammatory cytokines such as MCP-1, IL-6, and IL-8. Since IL-8 plays a major role in immunopathology caused by *H. pylori*, we describe a general protocol for the measurement of this cytokine (see Note 1).

In addition to production of proinflammatory cytokines, *H. pylori* infection also leads to IFNβ expression (8). Measuring IFNβ production in human cells presents a challenge. Commercially available IFNβ ELISA kits are usually very expensive and time-consuming. Luciferase reporter systems in Human Embryonic Kidney (HEK) cells provide an alternative method of inexpensive quantification of IFNβ expression along with the convenience of ectopic expression of proteins of interest. The reporter assay can also be modified to measure activation of other pathways activated by *H. pylori* such as NF-κB, JNK, and AP-1. HEK cells are also strong producers of IL-8, a cytokine strongly stimulated by *H. pylori* infection (10). MEF cells isolated from mice of various genotypes can also be used to study in vitro infections of *H. pylori*. MEFs are capable of producing a wide range of cytokines (6).

The early inflammation triggered by activation of TLRs and Nod1 leads to priming of the adaptive immune response, resulting in the production of antibodies targeted to *H. pylori*.

1. Note

1. IL-8 is a chemokine produced by human but not mouse cells. MCP-1 and IL-6 are produced by both human and mouse cells.

Acknowledgments

The authors thank Glennice Ryan, Anna Cerny, Melvin Chan, Michael King, An Zacharia, and Shenghua Zhou for their valuable input.

References

1. Peek RM Jr, Fiske C, Wilson KT (2010) Role of innate immunity in Helicobacter pylori-induced gastric malignancy. Physiol Rev 90:831–858
2. Saleh M, Trinchieri G (2011) Innate immune mechanisms of colitis and colitis-associated colorectal cancer. Nat Rev Immunol 11:9–20
3. Mandell L, Moran AP, Cocchiarella A, Houghton J, Taylor N, Fox JG, Wang TC, Kurt-Jones EA (2004) Intact gram-negative Helicobacter pylori, Helicobacter felis, and Helicobacter hepaticus bacteria activate innate immunity via toll-like receptor 2 but not toll-like receptor 4. Infect Immun 72:6446–6454
4. Fox JG, Rogers AB, Whary MT, Ge Z, Ohtani M, Jones EK, Wang TC (2007) Accelerated progression of gastritis to dysplasia in the pyloric antrum of TFF2 -/- C57BL6×Sv129 Helicobacter pylori-infected mice. Am J Pathol 171:1520–1528
5. Kurt-Jones EA, Cao L, Sandor F, Rogers AB, Whary MT, Nambiar PR, Cerny A, Bowen G, Yan J, Takaishi S, Chi AL, Reed G, Houghton J, Fox JG, Wang TC (2007) Trefoil family factor 2 is expressed in murine gastric and immune cells and controls both gastrointestinal inflammation and systemic immune responses. Infect Immun 75:471–480
6. Kurt-Jones EA, Sandor F, Ortiz Y, Bowen GN, Counter SL, Wang TC, Finberg RW (2004) Use of murine embryonic fibroblasts to define Toll-like receptor activation and specificity. J Endotoxin Res 10:419–424
7. Kawai T, Akira S (2010) The role of pattern-recognition receptors in innate immunity: update on Toll-like receptors. Nat Immunol 11:373–384
8. Watanabe T, Asano N, Fichtner-Feigl S, Gorelick PL, Tsuji Y, Matsumoto Y, Chiba T, Fuss IJ, Kitani A, Strober W (2010) NOD1 contributes to mouse host defense against Helicobacter pylori via induction of type I IFN and activation of the ISGF3 signaling pathway. J Clin Invest 120:1645–1662
9. Polk DB, Peek RM Jr (2010) Helicobacter pylori: gastric cancer and beyond. Nat Rev Cancer 10:403–414
10. Torok AM, Bouton AH, Goldberg JB (2005) Helicobacter pylori induces interleukin-8 secretion by Toll-like receptor 2- and Toll-like receptor 5-dependent and -independent pathways. Infect Immun 73:1523–1531
11. Takenaka R, Yokota K, Ayada K, Mizuno M, Zhao Y, Fujinami Y, Lin SN, Toyokawa T, Okada H, Shiratori Y, Oguma K (2004) Helicobacter pylori heat-shock protein 60 induces inflammatory responses through the Toll-like receptor-triggered pathway in cultured human gastric epithelial cells. Microbiology 150:3913–3922

Chapter 24

Methods for In Vivo and In Vitro Analysis of Innate Immune Responses to *Helicobacter pylori* Infection

Milan K. Patel, Glennice N. Ryan, Anna M. Cerny, and Evelyn A. Kurt-Jones

Abstract

It is estimated that half of the world's population is infected by *Helicobacter pylori* (*H. pylori*) (Polk and Peek, Nat Rev Cancer 10:403–414, 2010; Peek et al., Physiol Rev 90:831–858, 2010). Following infection, *H. pylori* induces a chronic innate immune response that is thought to contribute to gastric complications. Due to the widespread prevalence of *H. pylori*, it is important to study the innate immune responses that result from the infection. A variety of in vitro and in vivo techniques have been developed by our laboratory to study this immune response (Fox et al., Am J Pathol 171:1520–1528, 2007; Kurt-Jones et al., Infect Immun 75:471–480, 2007; Kurt-Jones et al., J Endotoxin Res 10:419–424, 2004). These methods are described here.

Key words: ELISA, Cytokine, Luciferase, NF-κB activation, TLRs, *H. pylori* bacteria

1. Introduction

Mice infected with *H. pylori* develop severe gastric immunopathology that parallels complications in humans. The immune response to Helicobacter infection has been well studied and is multifaceted (1–11). We have developed numerous methods to study in vivo *H. pylori* infections ranging from isolation of infected tissues to cytokine analysis. Here we present several protocols to measure the Helicobacter specific immune response.

JeanMarie Houghton (ed.), *Helicobacter Species: Methods and Protocols*, Methods in Molecular Biology, vol. 921,
DOI 10.1007/978-1-62703-005-2_24,

2. Materials

2.1. Animal Organ Processing and Cytokine Analysis from Tissues

1. Autoclaved/sterilized surgical instruments for animal dissection.
2. Isoflurane.
3. 70% Ethanol.
4. Phosphate Buffered Saline (PBS).
5. Solution: 4% paraformaldehyde/0.1% glutaraldehyde.
6. Complete Protease-inhibitor tablet.

2.2. Development and Culturing of Stably-Transfected HEK293 Cell Lines

2.2.1. Stable Transfection of HEK293 Cells

1. HEK293 cells (ATCC, Cat# CRL-1573).
2. Plasmids encoding puromycin resistance or hygromycin resistance.
3. Plasmids encoding human CD14, TLR2 or TLR4. TLRs can be FLAG-epitope tagged at the N-terminus or GFP-tagged at the C-terminus.
4. Puromycin and hygromycin.
5. Transfection reagent (we recommend GeneJuice reagent from EMD Chemicals).
6. Anti-FLAG mAb, anti-CD14 mAb, PE-labeled goat anti-mouse IgG.
7. FACS machine.
8. Freezing medium: 90% FCS (HyClone) and 10% DMSO.
9. Cryovials (1.5 ml).

2.2.2. Cell Culture of Stably Transfected HEK293 Cells

1. Dulbecco's modified Eagle's medium (DMEM) supplemented with 10% fetal bovine serum, and Penicillin/Streptomycin (Pen/Strep).
2. PBS.
3. Selection Media:
   (a) DMEM containing puromycin (5 μg/ml) and hygromycin (200 μg/ml).
4. Optional antibiotic additive: Ciprofloxacin (10 μg/ml) in PBS.
5. Trypsin-Versene mixture.

2.3. Isolation and Culturing of Primary Cell Lines

2.3.1. Isolation of Macrophage Cell Lines

1. WT or knockout mice (6–12 weeks of age) for isolation of macrophages.
2. 4% Thioglycollate solution. The solution should be prepared at least 4 weeks prior to use. After dissolving powder, the solution is autoclaved and stored at room temperature.
3. Freezing medium: 90% FCS and 10% DMSO.
4. Cryovials (1.5 ml).

2.3.2. Isolation of Peripheral Blood Mononuclear Cells and Human Monocytes from Whole Blood or Leukopacks

1. Human peripheral blood or leukopacks anticoagulated with heparin.
2. 50 ml Conical tubes.
3. Ficoll-Hypaque gradients.
4. Centrifuge for conical tubes.
5. RPMI 10% FBS.
6. RPMI 0% FBS (for leukopacks).
7. Anti-CD2 and anti-CD3 mAb.
8. Goat anti-mouse coupled magnetic beads.
9. Phycoerythrin-conjugated antibodies specific for CD14, CD4, and CD3.
10. FACS machine.
11. Freezing medium: 90% FCS and 10% DMSO.
12. Monocyte media: RPMI-1640 plus 10% heat-inactivated fetal calf serum (FCS) and 1% Pen/Strep.

2.3.3. Isolation of MEF Cell Lines

1. WT or knockout mice with timed pregnancies for isolation of murine embryonic fibroblast (MEF) cell lines.
2. Sterilized (autoclaved) dissecting instruments (fine scissors and tweezers).
3. 15 ml conical tubes.
4. Ice-cold Trypsin (we suggest GIBCO, 0.25% trypsin containing EDTA).
5. 10 cm Tissue culture-grade plates.
6. PBS w/o cations.
7. 37°C Water bath.
8. DMEM containing 10% HyClone FCS, 1% L-glutamine, 1% Pen/Strep.
9. Cell scraper.
10. Centrifuge for conical tubes.
11. Freezing medium: 90% FCS (HyClone) and 10% DMSO.
12. Media with some additional nutrients (15% FCS, 1% of nonessential amino acid/HEPES buffer/sodium pyruvate).
13. Cryovials (1.5 ml).
14. Large mouth pipette.
15. T-25 Flask.

2.4. Stimulation of Cells for Cytokine Analysis and RNA Extraction

1. Cells (HEK293 cells transfected with TLRs, macrophages, peripheral blood mononuclear cells (PBMCs), monocytes, and MEFs).
2. Control stimulants: Zymosan, Peptidoglycan, LPS, human IL-1β, PMA, Pam_2CSK_4, Pam_3CSK_4.

3. 24-Well tissue-culture grade plates.
4. Stimulated or non-stimulated cells collected using trypsin-EDTA and scraping with a rubber policeman followed by centrifugation to pellet the cells. RNA lysis buffer is added to the washed cell pellet. Alternatively, RNA can be extracted directly from adherent cells in the culture plate by adding RNA lysis buffer to the washed monolayer. Tissues require processing with tissue shredder.
5. RNA extraction kit (we suggest Qiagen-RNeasy Kit™).
6. Reverse transcription kit (we suggest Qiagen-One Step RT-PCR Kit™).
7. Thermal Cycler.

2.5. Measuring Cytokine Secretion by ELISA

1. Supernatants collected from stimulated and unstimulated cells transferred to 24-well tissue culture-grade plates and stored at –20°C until use.
2. ELISA assay kit (Pierce/Endogen IL-8 ELISA or Pharmingen OptEIA IL-8 kit).
3. ELISA plate reader (Dynatech MR7000; Dynatech Laboratories, Inc., Chantilly, VA).

2.6. Luciferase Assay

1. Cells (HEK293 expressing TLRs and transfected with NF-κB Firefly-Luciferase reporter plasmid and Renilla-Luciferase reporter to normalize for transfection efficiency).
2. Transfection Reagents (see Note 1).
3. 96-Well tissue culture-grade plates.
4. 96-Well white luminometer plates.
5. Dual-luciferase kit (Promega, Dual Glo Luciferase Assay System).
6. Luminometer.

2.7. Enzyme-Linked Immunosorbent Assay for Serum IgG2c and IgG1 Responses to H. Pylori

1. Immulon™ II plates (Thermo Labsystems) coated with *H. pylori* cell wall components.
2. Biotinylated secondary antibodies including monoclonal anti-mouse antibodies produced by clones A85-1 and 5.7 (Pharmingen-BD Biosciences, San Jose, CA) for detecting IgG1 and IgG2c, respectively.
3. Extravidin peroxidase.
4. 2,2′-azino-bis (3-ethylbenzthiazoline-6-sulfonic acid) diammonium salt (ABTS) substrate for color development.
5. Wash buffer: PBS + 0.01% Tween-20.
6. ELISA plate reader.

3. Methods

3.1. Animal Organ Processing

Helicobacter pylori infection is best examined using a combination of both in vivo and in vitro systems. The following method has been optimized for examining tissues isolated from an *H. pylori* infected mouse for cytokine production, RNA, and histology use. All animal experiments must be conducted according to an approved IACUC protocol.

3.1.1. Dissection of the Animal

1. Humanely euthanize the animal (e.g., isoflurane) and exsanguinate the animal by severing the axillary vein.
2. Disinfect the abdomen with 95% then 70% ethanol.
3. Make a longitudinal incision to open the abdominal cavity.
4. Dissect the stomach and intestine using scissors and tweezers.
5. Separate the stomach from the animal by cutting from the duodenum and esophagus.
6. Once removed from the body, cut the stomach along the greater curvature.
7. Holding the stomach with tweezers, wash the stomach in 1× PBS.

3.1.2. Processing of the Stomach

1. Place the stomach flat (this is best accomplished by either pinning, or adhering the tissue to a histology sponge) and opened in order to cut into three longitudinal sections. A fresh sharp sterile razor blade works best for this.
2. Process either the left or right side for cytokine analysis. Be consistent between samples. Place this sample in 500 μl of 1× PBS containing protease inhibitor cocktail (1 complete mini tablet into 14 ml of 1× PBS).
3. Place the other sample into a microcentrifuge tube and flash freeze in a dry ice/70% ethanol bath. This sample will be used for RNA analysis. Store at -80°C.
4. The middle section is bisected from the esophagus to the pylorus and fixed as two pieces by placing the tissue into a cassette and placed into a 4% paraformaldehyde/0.1% glutaraldehyde solution for 4–6 h at 4°C. After 4–6 h the fix is replaced with 70% ethanol and tissues can be processed for histological analysis.

   See Fig. 1 for details of how the tissue is sectioned.

3.1.3. Cytokine Analysis of Tissue Samples

1. Homogenize the samples for cytokine analysis using a handheld electric homogenizer (100-132-137, Branson Ultrasonics Corporation).
2. Centrifuge the samples at 17,000–18,000 ×*g* for 10 min.

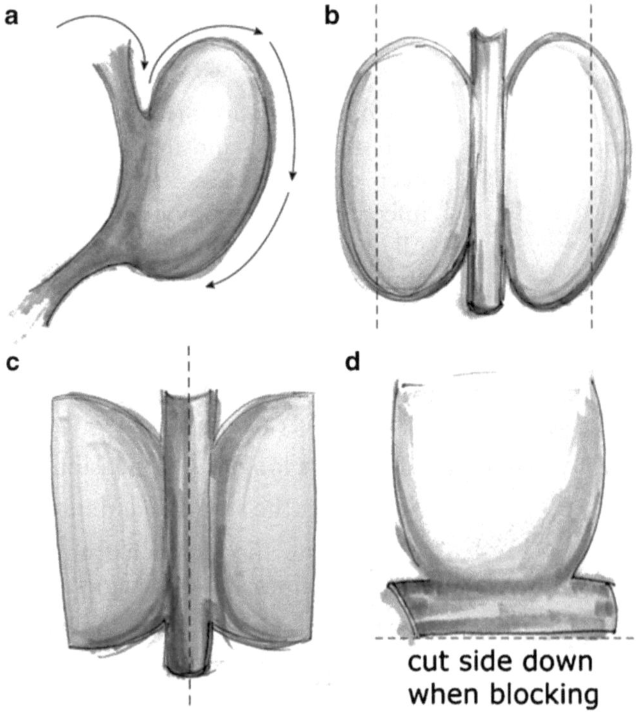

Fig. 1. Orientation of stomach tissue for processing. (**a**) The stomach is opened along the greater curvature (*arrow*) and unfolded like a butterfly. (**b**) The edges of the "wings" are removed and processed for RNA and cytokines. (**c**) The remaining piece is bisected from the esophagus to the pylorus and processed so that the middle edge is sectioned first (**d**).

3. At this point, samples can be stored at –80°C until cytokine analysis is performed.
4. Prepare a 1:5 dilution of the sample (e.g., 22 μl of sample in 88 μl of 1× PBS) to run an ELISA for cytokine analysis (e.g. mIL-6, mMCP-1, mRANTES).

3.2. Development and Culturing of Stably-Transfected HEK293 Cell Lines

Stably transfected HEK293 cell lines expressing FLAG-epitope tagged or GFP-tagged TLR2, TLR4, and CD14 are used to determine the role of these receptor molecules in the innate immune response to *H. pylori* infection.

3.2.1. Stable Transfection of HEK293 Cell Lines

1. HEK293 cells are co-transfected with TLR plasmids and an antibiotic resistance gene (Puro® or Hygro®) at a 10:1 ratio of TLR plasmid to antibiotic resistance gene. It is important to establish the sensitivity of your HEK cells to the antibiotic within a week or two of setting up the transfection as the dosage necessary to kill the non-transfected cells can vary over time.
   (a) Determine the kill curve for HEK293 cells in puromycin (or hygromycin) selection. Plate the cells at the same density as you would for a transfection.

(b) The following day add selection media in the following range (puromycin 0.5–10 μg/ml and hygromycin 0.1–0.4 mg/ml).

(c) Twenty-four hours later examine the killing efficiency in the selection media the dose that kills >95% of puromycin (or hygromycin) negative cells after 24 h is the concentration to use for selection (see Note 2).

2. Transfect HEK293 cells with plasmids encoding puromycin (or hygromycin) resistance together with plasmids encoding N-terminal Flag-epitope tagged human TLR2 or TLR4 and/or human CD14 using GeneJuice reagent according to the manufacturer's protocol. Use a molar ratio of 10:1, TLR plasmid:Puro® or Hygro®.
3. Forty-eight hours later, add the pre-determined amount of puromycin (or hygromycin) to the cultures. We generally add 2× the minimal effective toxic concentration of antibiotic.
4. Change media daily to remove dead cells until puromycin (or hygromycin) resistant cells are observed (usually within 48 h). Resistant cells will expand over 7–14 days (Control cells that did not receive Puro® or Hygro® plasmid should all die within 7 days).
5. Isolate clones of puromycin (or hygromycin) resistant cells. Expand and freeze down stocks in freezing medium containing 90% FCS and 10% DMSO.
6. Analyze the surface expression of proteins using anti-FLAG mAb to detect TLR proteins and anti-CD14 mAb followed by a PE-labeled goat anti-mouse IgG by FACS for surface expression of proteins.
7. Select clones expressing equivalent levels of TLR proteins by Western blot.

3.2.2. Cell Culture of Stably Transfected HEK293 Cells

For in vitro analysis of *H. pylori* infection, our lab has generated a panel of TLR-expressing stable-cell lines. In particular, HEK cells stably transfected with TLR2 and TLR4 have been very useful for studying NF-κB responses induced by *H. pylori*. HEK cells are very good producers of IL-8, an NF-κB dependent cytokine. In addition, HEK cells are readily transfected allowing ectopic expression of TLR genes or reporter plasmids to study how the different components of *H. pylori* bacteria activate innate immunity. Cells should be maintained in selection media.

1. Optimal activation of HEK cells occurs at 80% confluence. Pass the HEK cells every 2–3 days based on cell density/confluence. Split the cells at 1:2 or 1:4 the day before setting up bacterial stimulation assays. This assures that the HEK cells are in log-phase growth at the time of analysis. Use brief exposure to trypsin-versene to detach cells for passage. Do not use PBS/

EDTA to detach cells since significant cell death will be observed with prolonged exposure to EDTA.

2. For maintaining the cells, passage at 1:5 or 1:10 every 48–72 h.
3. After HEK cells are detached with trypsin, there is a 12 h lag while the cells recover. Thereafter, cells will double every 18–30 h depending on the individual cell line (see below).
4. Plating HEKs based on density/confluence: For surface area we assume that a T75 flask is roughly equal to a 24-well plate, so a confluent T75 flask is sufficient for two 24-well plates at 50% confluence.
5. After the lag phase both the HEK and the TLR2/CD14 cells have about an 18 h doubling time.
6. The TLR4/MD2 cells have about a 30–36 h doubling time. It is very important to use an endotoxin-free FCS for culturing TLR4 expressing cells. Low levels of endotoxin contamination will lead to chronic activation of the TLR4 cells with high "spontaneous" IL-8 secretion and NF-κB activation. Endotoxin contaminated FCS will ultimately kill the cells (see Note 3).
7. The HEK293 stably-transfected cell lines grow at different rates, TLR4 transfectants grow more slowly than TLR2 transfectants, so adjust the number of cells to achieve ~80% confluence on the day that stimulants are added (see Note 4).

3.3. Isolation and Culturing of Primary Cell Lines

For all animal experiments, please assure that all experiments are performed under an approved IACUC protocol.

3.3.1. Isolation of Macrophage Cell Lines

1. For elicited macrophages, administer 1 ml of 4% thioglycollate intraperitoneally, and harvest the cells 4 days later. Recovery of cells is generally about 10^6 per mouse of resident macrophages. You can expect $1–2 \times 10^7$ macrophages on day 4 following thioglycollate treatment.
2. Harvest peritoneal exudate cells (PECs) by peritoneal lavage of WT or knockout mice. Mice are humanely sacrificed (for example, with CO_2 followed by cervical dislocation). The peritoneum is exposed and 10 ml of sterile saline is injected into the peritoneum. The needle is withdrawn and the mouse gently shaken or massaged to suspend the PECs. A syringe fitted with an 18 guage needle is inserted along the midline and the needle lifted gently to tent the peritoneal membrane. Fluid is collected by slowly pulling back on the syringe. Generally 8 ml of fluid can be recovered.

3.3.2. Isolation of PBMCs and Human Monocytes from Whole Blood

Human peripheral blood is an excellent source of innate immune cells, particularly monocytes.

All experiments using human cells and/or human tissues must be conducted under an approved IRB protocol.

1. Collect 30 ml of whole heparinized blood and transfer into a 50 ml centrifuge tube.
2. Gently underlay the blood with 12 ml of ficoll/lymphocyte separation media. This is most easily accomplished by filling a pipet with ficoll solution, then lowering the filled pipet into the tube containing the blood and releasing the seal once the ficoll-filled pipet is sitting on the bottom of the tube. The blood layer will lift up and float on the ficoll layer as the pipet empties. Once the pipet has stopped emptying, place your finger over the top of the pipet (to prevent dripping) and slowly withdraw the pipet from the tube. If done correctly there will be a sharp interface between the clear ficoll layer on the bottom of the tube and the blood layer above it.
3. Spin sample at 800 × *g* for 40 min at 4°C with no brake on the centrifuge. This avoids sudden jolting of the tube which can cause the layers to mix.
4. Harvest cells from the interface between the blood and ficoll layers. Try to harvest cells from the top of the ficoll layer taking care to avoid harvesting the clear ficoll itself.
5. Wash the harvested cells by adding RPMI 10% FBS and centrifuging at 400 × *g*, 10 min. Repeat wash step to remove all of the ficoll.
6. Resuspend washed cells in a known volume of medium and count. These are PBMCs and contain a mixture of monocytes, lymphocytes and NK cells.
7. PBMC can also be harvested from leukopacks. For Leukopacks, dilute 1:1 volume/volume with RPMI 10% FBS, then add 30 ml of diluted leukopack medium mixture to centrifuge tube. Underlayer with ficoll and proceed as above.
8. To further isolate monocytes from the PBMCs perform magnetic bead selection. Lymphocytes, including T cells and NK cells, are depleted by incubating the PBMC with anti-CD2 and anti-CD3 monoclonal antibodies followed by goat anti-mouse coupled magnetic beads.
9. After removal of the T cells and NK cells with a magnet, the purity of the monocyte preparation is determined by flow cytometry with phycoerythrin-conjugated antibodies specific for CD14 (monocyte marker), CD4 and CD3 (lymphocyte markers). Monocyte purity is routinely >90% after lymphocytes and NK cell depletion. Alternatively, monocytes can be positively selected using anti-CD14 mAb and magnetic beads. Magnetic kits are available from Miltenyi for both positive and negative selection of monocytes.
10. Monocytes are cultured in RPMI-1640 medium supplemented with 10% heat-inactivated fetal calf serum (FCS) and 1% penicillin-streptomycin at a density of 10^5 per ml in tissue culture dishes.

3.3.3. Culturing MEF Cell Lines

MEFs are primary cells with a limited lifespan in culture. Our laboratory has found that MEFs are an excellent in vitro system for studying the innate immune response (5). They express most of the NLRs and TLRs. They can produce a wide-range of cytokines including IL-1β, TNFα, IL-6, MCP-1, and Rantes (*Note*: MEFs do not produce IL-8). In addition, MEFs can be isolated from knockout mice to study the effect of individual genes of interest on the innate immune response. In addition to cytokine production, MEFs are also very good producers of the type I IFNs.

1. Set up 2 × 10 cm dishes containing sterile 1× PBS w/o cations. Sacrifice the pregnant female at 14–16 d.p.c. using CO_2 or isoflurane followed by cervical dislocation. Dissect out the uterine horns, place into the 10 cm dish.
2. Separate each embryo from its placenta and surrounding membranes. Place the embryo in the first 10 cm dish and remove the head for genotyping. Cut away and discard dark red internal organs, especially liver, to avoid non-fibroblast contaminations. Move the embryo to the second 10 cm dish and rinse to remove small loose pieces of debris and as much blood as possible. Transfer the washed embryo to a 15 ml conical containing 5 ml ice cold trypsin solution.
3. Incubate embryo in trypsin solution at 4°C overnight to allow diffusion into the embryo.
4. Activate trypsin by incubation for 15 min at 37°C.
5. Pour off the excess trypsin solution.
6. Add 4 ml DMEM/FCS (DMEM with 10% HyClone FCS, 1% L-glu, 1% Pen-Strep) and mince specimen by pipetting up and down using a large bore pipette. Only small clumps should remain. Let debris settle by gravity.
7. Plate out the digested and minced cell solution in a 10 cm dish (this is "passage No. 0") and add an additional 4 ml of DMEM. Allow fibroblast cells to attach overnight. Change the medium the following day, and split cultures as soon as they become confluent.
8. Splitting: Wash the cells with 1× PBS w/o cations twice. Clumps and debris must be washed off with PBS. Add 4 ml of 1× PBS w/o cations and scrape cells off the plate with a cell scraper (or use trypsin). Replate 1/4 to 1/5 of cells into a new dish with 8 ml of DMEM. This is equivalent to splitting the cells at an approximate density of 10^4 cells/cm^2 every 3 days.
9. Freeze down cells at each passage. Transfer the harvested cells to a 15 ml conical tube and pellet cells by centrifuging for 4 min at 300 × *g*. Discard media, add 1 ml of freezing medium

(HyClone FCS w/10% DMSO) to cell pellets. Resuspend cells in the freezing medium and transfer to cryovial (1.5 ml). Store cells at −80°C for 1–2 weeks. Then transfer to liquid nitrogen for long term storage.

10. Freeze down cells at passage 1, 2, 3 and 4 (see Note 5).
11. Thawing MEFs: Swirl cryovial in 37°C water bath immediately after removing the cryovial from the liquid nitrogen tank. Allow half of the amount frozen to thaw. Spray the outside of the tube with 70% ethanol prior to opening vial.
12. Add 1 ml of DMEM/10% FCS media to the cell solution in the cryovial and gently pipette up and down to thaw the remaining frozen cells. Transfer thawed cells to T25 tissue culture flask. Add medium to bring volume to 5 ml. After incubating overnight at 37°C, remove medium and replace with fresh medium. MEFs can be grown to passage 7–8 if passed every 3 days at density of 10^4 cells/cm^2. After passage 8, MEFs will become senescent and cultures are discarded.
13. If MEFs don't appear to grow well the next day immediately after thaw, discard half of the old media and add new media with some additional nutrients (15% FCS, 1% of nonessential amino acid/HEPES buffer/sodium pyruvate).
14. Monitor MEFs very carefully while in culture. If they begin to elongate and/or the doubling time increases significantly indicating senescence, they are beyond their useful passage number. Avoid overgrowing the cultures or allowing the cells to become confluent which can result in early senescence. When MEFs are properly grown it is easy to expand these cells from passage 2–7 with very little loss of cell viability and integrity.

3.4. Stimulation of Cells

3.4.1. Stimulation of HEK293 Cells

1. Plate HEK293 cells in tissue culture dishes and allow them to adhere and grow for a day in culture until they are about 80% confluent.
2. Optimal cultures contain HEK293 cells at $2–5 \times 10^4$ cells/well in 24-well plates containing RPMI 1640 plus 10% heat-inactivated fetal calf serum (see Note 6).
3. Cells in 24-well plates are cultured in a final volume of 1 ml medium. It is often convenient to plate the HEK cells in 500 μl medium, grow overnight, then add 500 μl of medium containing stimulants or 500 μl medium alone to bring the final volume to 1,000 μl.
4. Stimulate 5×10^4 HEK cells with varying doses of bacteria, i.e., *H. pylori* ($10^5–10^7$ cfu/well) or with control stimulants such as LPS (TLR4 ligand, 100 ng/ml). Also include wells treated with medium alone, Pam_3CSK_4 (TLR2 ligand), and IL-1β (non-TLR control ligand). These control wells ensure that the cells are healthy and respond to appropriate stimulants. These

controls may be useful if you want to normalize the responses of different cell lines.

5. Culture cells with stimulants for 18–24 h at 37°C. Collect supernatants (about 750 μl) and transfer to fresh 24-well tissue culture plate(s). Supernatants can be stored frozen at −20°C until you are ready to set up the ELISA. HEKs secrete IL-8 and can be used for NF-κB luciferase assays.

3.4.2. Stimulation of Macrophages (PECs)

1. Culture PECs at 10^6 cells/well in 24-well plates in RPMI 1640 plus 10% heat-inactivated fetal calf serum (see Note 7).
2. Incubate 1–2 h after plating then wash with 1× PBS.
3. Add media containing *H. pylori* (10^5–10^7 cfu/well) or LPS (100 ng/ml) as a control. Stimulate 16–24 h at 37°C.
4. Harvest culture supernatants after stimulation, and proceed to ELISA or freeze at −20°C for future analysis.
5. Murine PECs make mouse IL-6, MCP-1, RANTES, and IFNβ.

3.4.3. Stimulation of PBMCs and Monocytes

1. Plate human PBMCs or monocytes at 10^5 cells/24 wells in DMEM + 10% FCS + 1% L-glut and 1% pen-strep.
2. Stimulate after 1 h of plating with *H. pylori* (10^5–10^7 cfu/well) or LPS (100 ng/ml) as a control. Let stimulation go overnight (16–24 h) at 37°C at 10% CO_2.
3. Remove supernatants (can be frozen at this time at −20°C for up to 6 months) and measure cytokines by ELISA.
4. Human PBMCs make IL-8 (high-levels), RANTES, and MCP-1. Interferon bioassays can also be performed on PBMCs.

3.4.4. Stimulation of MEFs

1. Plate MEFs at 10^5 cells/well in DMEM + 10% FCS + 1% L-glut and 1% pen-strep.
2. Stimulate after 1 h of plating with *H. pylori* (10^5–10^7 cfu/well) or LPS (100 ng/ml) as a control. Let stimulation go overnight (16–24 h) at 37°C.
3. Remove supernatants (can be frozen at this time at −20°C for up to 6 months) and measure cytokines by ELISA.
4. MEFs make: IL-6, MCP-1, RANTES, IFNα, and IFNβ.

3.5. Measuring Cytokine Secretion by ELISA

Following *H.pylori* infection, the innate immune response leads to the production of several cytokines. IL-8 is an example of one of several cytokines that are released in response to *H. pylori* infection. Cytokine levels in culture supernatants or in tissue extracts can be measured using specific cytokine ELISA assays.

1. Plate HEK293 cells in 24-well culture dishes at 40–50% confluence.

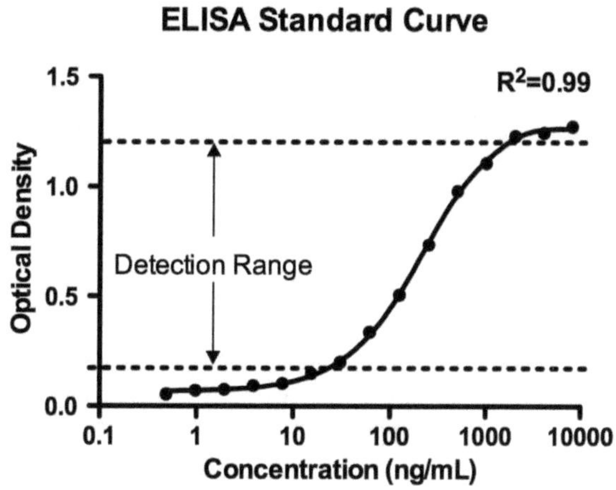

Fig. 2. ELISA Standard Curve. The figure shows a typical ELISA curve generated from the assays. A good assay will have an R^2 value of 0.99 or greater. The *y*-axis is the Optical Density which is on a linear scale, while the *x*-axis is the concentration in ng/ml which is on a log scale. The linear range falls only in the area between the dashed lines. Readings that fall below or above this region are not accurate to determine concentration measurements.

2. 24 h Later (cells should be 80% confluent) replace the culture medium with fresh medium ± stimulant.
3. After 18 h of stimulation, harvest the culture supernatants. Collect 0.5 ml from each 1 ml culture, and transfer it to a new 24-well plate.
4. Store supernatants by sealing the plate with parafilm and freezing at −20°C until ready to set up the ELISA.
5. IL-8 levels in the supernatants are measured using a Pierce/Endogen IL-8 ELISA assay kit (see Note 8). It is important to create a proper standard curve as described below and shown in Fig. 2.
6. It is recommended to have at least 12 different concentrations to create a standard curve. The standard curve delineates the allowed range of detection for the samples. The OD values of the samples must lie within the linear range of the standard curve. OD values above or below the linear range (shown as values outside of the dotted lines in Fig. 2) are outside the limit of detection and may lead to irreproducible results. It is also important to make sure that the value of the correlation coefficient (R^2) is close to 1 for the standard curve (see Note 9).
7. For readings that fall within the linear range, the final concentration is determined by multiplying the readings by the dilution factor. If the readings do not fall between the linear range limits, it is important to repeat the ELISA with the supernatants diluted accordingly, i.e., if the value is at the upper limit of detection the supernatants should be run at a greater dilution.

Values falling within the linear range are used to determine the amount of cytokine based on the standard. This number is multiplied by the dilution factor to determine the actual concentration of the cytokine in the sample.

8. Other cytokines such as IL-6, MCP-1, and TNF-alpha can also be measured using an ELISA. Our laboratory has found that a 1:5 dilution of supernatant is best for IL-6 and MCP-1, while undiluted supernatant should be used for TNF-alpha.

3.6. Luciferase Assay

Luciferase reporter assays are an inexpensive way to measure activation of various signaling pathways important to the innate immune response. Our laboratory has developed robust luciferase assays to determine NF-κB responses, IFNβ response, and AP1 response following *H. pylori* infection. Usually, HEK cells are used for luciferase assays since they are readily transfected. However, the technique can be easily applied to other cell types.

3.6.1. Transfection Protocol

1. Plate HEK293 cells the day before in 96-well plates in 200 μl/well. Cells should be less than 50% confluent the next day for transfection. For HEK cells use 20,000–25,000 cells/well.
2. Prepare the mixture of DNA and the transfection solution (i.e., GeneJuice) in V-bottom 96-well plates. This allows direct transfer of the DNA/GeneJuice mix to the 96-well plate with cells. Each mixture will contain the NF-κB firefly-luciferase reporter plasmid along with a renilla-luciferase plasmid under the control of a constitutively active promoter. The Renilla luciferase levels serve as an internal standard to normalize for transfection efficiency.
3. Use clean, endotoxin-free DNA. Determine the concentration using a spectrophotometer. Generally concentrations of 0.1–1.0 μg/ml are the best (For HEK cells the maximum amount of DNA is 300 ng/well).
4. For Luciferase assay use the following amounts of DNA:

Assay	Reporter (concentration)	Control (concentration)
AP1	AP1-firefly (80–100 ng/well)	EF1α or SV40 Renilla (20–40 ng/well)
NFκB	NFκB-firefly (80–100 ng/well)	EF1α or SV40 Renilla (20–40 ng/well)
IFNβ	P125-firefly (80–100 ng/well)	EF1α or SV40 Renilla (20–40 ng/well)

5. The choice of internal control is important. For the NF-κB luciferase assays use a promoter which is NF-κB independent (see Note 10).

6. Make up GeneJuice solution. For HEK 293 cells dilute 0.8 μl of GeneJuice with 9.2 μl of serum free media. This is enough for one well.
7. Mix equal volumes of diluted GeneJuice and DNA solution. (10 μl diluted GeneJuice + 10 μl of DNA mix). Incubate at room temperature for 15 min.
8. Add 20 μl DNA/GeneJuice mix per well to HEK cells. Incubate 37°C overnight.
9. The next day add *H. pylori* or control stimulants or medium alone to the cultures. Incubate for 3–6 h to induce luciferase reporter expression and then harvest cells.

3.6.2. Luciferase Assay Using the Promega Dual Stop and Glo Luciferase Assay Kit

1. HEK cells are adherent. Remove medium from HEK cultures. Pat plate dry on tissue and then add 50 μl/well of passive lysis buffer. Shake 15 min on rocking platform table to facilitate cell lysis.
2. Transfer 40 μl of cell lysate to white 96-well luminometer plates.
3. Add 20 μl of firefly luciferase assay reagent, wait for 10–15 min and read the relative light values on the luminometer.
4. Add 20 μl of renilla luciferase assay reagent, wait for 10–15 min to quench the firefly luciferase activity and read the renilla light values on the luminometer.
5. Calculate Firefly/Renilla values. For best results, use three replicates to be able to determine the standard deviation (Fig. 3).

4. Notes

1. (FuGene, GeneJuice). We found that these reagents give high transfection efficiency with minimal toxicity to the cells.
2. For hygromycin selection, killing takes about 72 h (our experience is that 200 μg/ml works best)
3. It is good to routinely test new lots of FCS before switching over TLR expressing cell lines. Generally, the higher quality FCS lots sold for hybridoma production are endotoxin-free, while cheaper FCS lots are sometimes contaminated. Use radiation sterilized tips and avoid autoclaved tips as these are sometimes endotoxin positive.
4. Aim to have all of the cells ready on the same day, thus, you need to plate about twice as many TLR4/MD2 cells as TLR2/

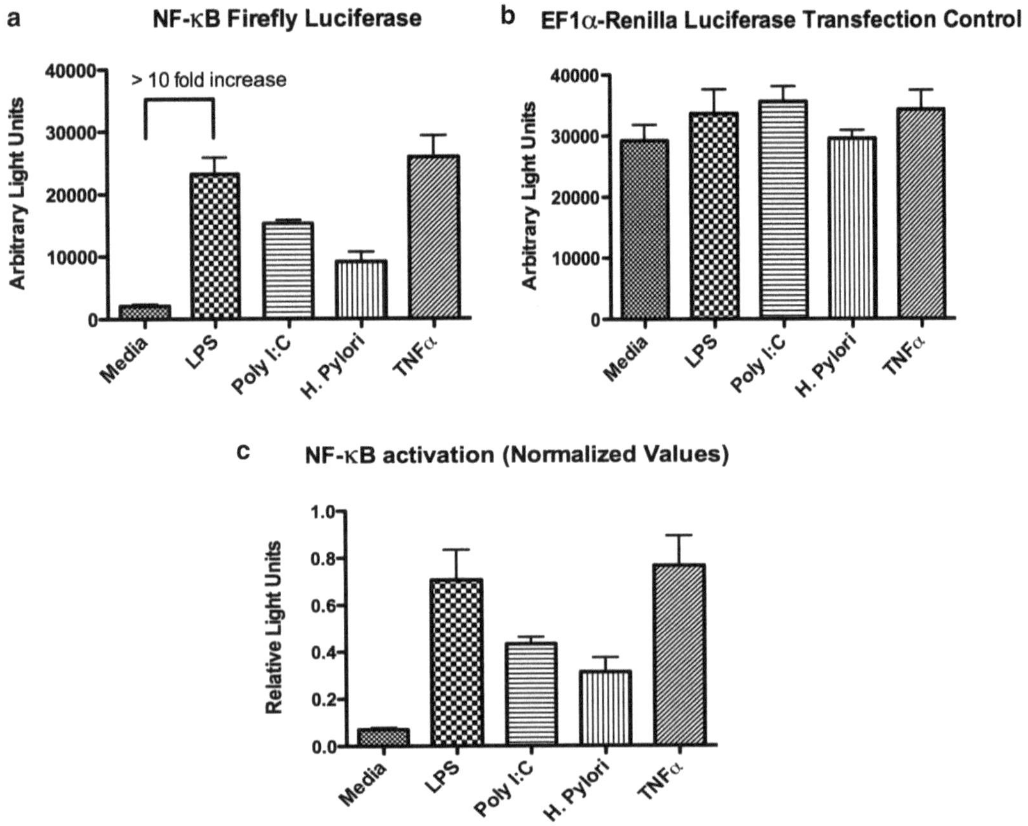

Fig. 3. NFκB luciferase assay. (**a**) The figure shows the readings from the firefly luciferase NFκB reporter assay using the following stimulants: lipopolysaccharide (LPS), Poly I:C, *H. pylori*, and TNFα with media as a negative control. The NFκB firefly reporter shows an inducible response to these stimulants. (**b**) The figure shows raw data from the EF1α promoter-renilla control for the same stimulants. The renilla reporter does not exhibit an inducible response to the stimulants. (**c**) The figure shows the relative luciferase values which were calculated as follows: Firefly/Renilla × 100.

CD14 or HEK cells OR you can plate all at the same density and assay the TLR2/CD14 and HEK cells 24 h after passage and assay the TLR4/MD2 cells 48 h after passage (e.g., split the TLR4/MD2 cells a day earlier than the HEKs or the TLR2/CD14 cells, so that the TLR4s have an extra day to grow).

5. Label: strain, mefs, passage (P1, P2, P3, etc.), date, initial.
6. Reminder: the growth rates of transfected HEK293 cells vary, therefore it is important to plate these cells at the appropriate density so that all cells will be 80% confluent upon stimulation.
7. When counting the cells remember to exclude the red blood cells (smaller than macrophages) OR you can lyse red bloods using red blood cell lysis buffer prior to counting.

8. We have also used Pharmingen OptEIA IL-8 kits.
9. Note that the standard concentrations are on a log-scale while the OD readings are on a linear scale.
10. Do not use a CMV-promoter driven renilla luciferase as an internal control. The CMV promoter is NF-kB sensitive and CMV-Renilla levels can fluctuate in a parallel with NF-κB levels. An SV40 or EF1α promoter driven renilla luciferase works well for normalizing NF-kB luciferase assays.

Acknowledgments

The authors thank Melvin Chan, Michael King, An Zacharia, and Shenghua Zhou for their valuable input.

We wish to thank Allycia Jones for artwork.

References

1. Polk DB, Peek RM Jr (2010) Helicobacter pylori: gastric cancer and beyond. Nat Rev Cancer 10:403–414
2. Peek RM Jr, Fiske C, Wilson KT (2010) Role of innate immunity in Helicobacter pylori-induced gastric malignancy. Physiol Rev 90:831–858
3. Fox JG, Rogers AB, Whary MT, Ge Z, Ohtani M, Jones EK, Wang TC (2007) Accelerated progression of gastritis to dysplasia in the pyloric antrum of TFF2 -/- C57BL6 x Sv129 Helicobacter pylori-infected mice. Am J Pathol 171:1520–1528
4. Kurt-Jones EA, Cao L, Sandor F, Rogers AB, Whary MT, Nambiar PR, Cerny A, Bowen G, Yan J, Takaishi S, Chi AL, Reed G, Houghton J, Fox JG, Wang TC (2007) Trefoil family factor 2 is expressed in murine gastric and immune cells and controls both gastrointestinal inflammation and systemic immune responses. Infect Immun 75:471–480
5. Kurt-Jones EA, Sandor F, Ortiz Y, Bowen GN, Counter SL, Wang TC, Finberg RW (2004) Use of murine embryonic fibroblasts to define Toll-like receptor activation and specificity. J Endotoxin Res 10:419–424
6. Saleh M, Trinchieri G (2011) Innate immune mechanisms of colitis and colitis-associated colorectal cancer. Nat Rev Immunol 11:9–20
7. Mandell L, Moran AP, Cocchiarella A, Houghton J, Taylor N, Fox JG, Wang TC, Kurt-Jones EA (2004) Intact gram-negative Helicobacter pylori, Helicobacter felis, and Helicobacter hepaticus bacteria activate innate immunity via toll-like receptor 2 but not toll-like receptor 4. Infect Immun 72:6446–6454
8. Kawai T, Akira S (2010) The role of pattern-recognition receptors in innate immunity: update on Toll-like receptors. Nat Immunol 11:373–384
9. Watanabe T, Asano N, Fichtner-Feigl S, Gorelick PL, Tsuji Y, Matsumoto Y, Chiba T, Fuss IJ, Kitani A, Strober W (2010) NOD1 contributes to mouse host defense against Helicobacter pylori via induction of type I IFN and activation of the ISGF3 signaling pathway. J Clin Invest 120:1645–1662
10. Torok AM, Bouton AH, Goldberg JB (2005) Helicobacter pylori induces interleukin-8 secretion by Toll-like receptor 2- and Toll-like receptor 5-dependent and -independent pathways. Infect Immun 73:1523–1531
11. Takenaka R, Yokota K, Ayada K, Mizuno M, Zhao Y, Fujinami Y, Lin SN, Toyokawa T, Okada H, Shiratori Y, Oguma K (2004) Helicobacter pylori heat-shock protein 60 induces inflammatory responses through the Toll-like receptor-triggered pathway in cultured human gastric epithelial cells. Microbiology 150:3913–3922

Chapter 25

Techniques for Following Labeled Cells In Vivo: Use of X/Y FISH, Techniques to Optimize Fluorescent Detection, and Beta-Galactosidase Detection

Michael Craig, Michael Schumacher, and Yana Zavros

Abstract

The redistribution and trafficking patterns of cells to different anatomic sites throughout the body is important during cancer development and metastasis. Interest in the origin and fate of gastric cancer stem cells has recently arisen, as it may explain the underlying mechanism of cancer development. The ability to monitor the migration patterns of cancer stem cells is imperative to understanding the functional changes associated with the migration and proliferation of these cells. Here we detail a collection of techniques that include fluorescent in vivo imaging, X/Y FISH, and beta-galactosidase detection that are used for following labeled cells in vivo after adoptive transfer or transplant of donor cells for identifying the migration and engraftment of donor cells within the recipient.

Key words: Gastric cancer, Stomach, Fluorescence membrane labeling, Inflammation, Cancer stem cells

1. Introduction

Methodologies enabling the trafficking of cells in vivo have made a tremendous impact on our understanding of the development of a number of gastric-related diseases including chronic inflammation and cancer. In particular, interest in the origin and fate of gastric cancer stem cells has recently arisen, as it may explain the underlying mechanism of cancer development (1–4). Cancer stem cell recruitment to the local tissue environment is a phenomenon that is related to the development of an inflammatory response in a variety of organs throughout the body (1, 3–7). In vivo approaches that investigators typically use to study migration and engraftment

JeanMarie Houghton (ed.), *Helicobacter Species: Methods and Protocols*, Methods in Molecular Biology, vol. 921,
DOI 10.1007/978-1-62703-005-2_25, © Springer Science+Business Media, LLC 2012

include implantation of cancer stem cells (either under the skin as a xenograft or "orthotopically" within the organ specific site) (8), adoptive transfer and bone marrow transplantation (1, 2). Determining the fate of cancer stem cells and understanding the functional changes associated with migration, engraftment, and proliferation require effective methods of identifying the in vivo trafficking pattern of these cells in the context of intact tissues and organ systems.

Understanding the origin and destination of gastric cancer stem cells and their link to the immune response advances our understanding of a variety of metastatic and inflammation-induced cancers. Such knowledge would facilitate the development of novel therapies for the treatment of gastric cancer. Here we detail a collection of techniques that are used for following labeled cells in vivo after adoptive transfer or transplant of donor cells. Techniques that include fluorescent in vivo imaging, X/Y fluorescence in situ hybridization (FISH), and beta-galactosidase detection have been shown to be valuable for identifying the migration and engraftment of donor cells within the recipient.

2. Materials

2.1. Fluorescence In Vivo Imaging of Cancer and Immune Cells

1. Phosphate buffered saline (PBS): For 1 l of 1× PBS: add 8 g of NaCl, 0.2 g of KCl, 1.44 g of Na_2HPO_4, and 0.24 g of KH_2PO_4 to 800 ml of distilled water. Adjust the pH to 7.4 with HCl and add distilled water to a total volume of 1 l.
2. RPMI-1640 Medium: HyClone RPMI-1640 media (Thermo Scientific) with sodium bicarbonate, supplemented with 0.3 g per liter glutamine, endotoxin and cell culture-tested and filter-sterilized.
3. RPMI-1640/1% fetal bovine serum: RPMI-1640 medium supplemental with fetal bovine serum.
4. CellVue® Kits for Membrane Labeling: Purchased from Polysciences, Inc.
5. Red blood cell lysis buffer: In 800 ml distilled water add 8.3 g, 1.0 g $KHCO_3$ and 1.8 ml of 5% EDTA. Filter-sterilize through 0.2 μm filter and make up to 1,000 ml with distilled water.

2.2. Beta-Galactosidase Detection

1. Phosphate buffered saline (PBS): For 1 l of 1× PBS: add 8 g of NaCl, 0.2 g of KCl, 1.44 g of Na_2HPO_4, and 0.24 g of KH_2PO_4 to 800 ml of distilled water. Adjust the pH to 7.4 with HCl and add distilled water to a total volume of 1 l.
2. 4% Paraformaldehyde: Measure 100 ml PBS into a glass beaker with a stir bar. Start stirring on hot plate IN THE HOOD.

Set hot plate setting to about 60°C while making sure the solution does NOT boil. Add 4 g of paraformaldehyde. Continue to heat until the liquid turns from cloudy to clear. The pH may need to be raised until the solution clears. This can be achieved by adding a few drops (2–3) of 1 N NaOH. When the solution is clear, the pH is corrected with HCl. Check the pH by pipetting a small amount of the solution onto a pH strip. If the pH is greater than 8 or below 7 keep titrating. DO NOT titrate on the pH meter as this will be damaged by the paraformaldehyde. Turn off heat and continue to stir until cool. Store at 4°C. Alternatively, ampoules of premade 10% paraformaldehyde may be purchased and diluted using 10 ml of 10× PBS, 80 ml of distilled water and 10 ml of paraformaldehyde from a freshly opened ampoule.

3. Rinse buffer: 0.1 M Na phosphate (pH 8.3–8.5), 2 mM $MgCl_2$, 0.1% Na deoxycholate, and 0.2% Triton X-100 (or Nonidet P-40).
4. X-gal staining solution: To 100 mg X-gal (Sigma Aldrich, B4252, 5-Bromo-4-chloro-3-indolyl-β-d-galactopyranoside) add 4 ml of *n*-dimethylformamide to make 25 mg/ml X-gal. Make stock solutions of the following:

   0.25 M Potassium ferricyanide (FW 329.26, $K_3Fe(CN)_6$): 8.23 g in 100 ml distilled water.

   0.25 M Potassium ferrocyanide (FW 422.39, $K_4Fe(CN)_6$): 10.56 g in 100 ml distilled water.

   To the 4 ml 25 mg/ml X-gal add 2 ml Potassium ferricyanide and 2 ml Potassium ferrocyanide and bring to 100 ml volume with the rinse buffer. Mix and use immediately.

2.3. Alternate Protocol to Fixing Tissue—Paraformaldehyde Perfusion

1. 1× PBS.
2. 4% paraformaldehyde, freshly prepared and chilled at 4°C.
3. Two syringes (20 ml) with 23 gauge needles.
4. Scissors, scalpel, forceps.

2.4. X/Y FISH

1. 20× Saline–sodium citrate (SSC) buffer: 173 g NaCl (3 M), 88.2 g (300 mM) sodium citrate. Adjust the pH to 7.0 with a few drops of 14 N solution of HCl. Make up to 1 l with water.
2. Hybridization buffer: 50%v/v formamide, 10% dextran sulfate, 2× SSC, and 0.5 M phosphate buffer pH 7.3.
3. Denaturing solution: 70%v/v formamide, 2× SSC. Adjust pH to 7.0.
4. Carnoy's fixative: 60 ml ethyl alcohol, 30 ml chloroform, and 10 ml acetic acid.

3. Methods

3.1. Fluorescence In Vivo Imaging of Cancer and Immune Cells

The following protocol has been optimized to follow cell migration in response to *H. pylori* infection in the stomach. The images shown in Fig. 1 demonstrate migration of splenocytes in vivo in response to *H. pylori* infection (see Note 1). Here we present a protocol to that may be used to record the trafficking pattern of immune cells to improve our understanding of how the immune response is initiated. This protocol may be modified to track the migration pattern of cancer cells in response to *H. pylori* infection (see Note 1). All animal work must be done under IACUC approval.

1. Remove spleen from a wild type uninfected mouse and store in ice-cold RPMI medium until homogenization.
2. Homogenize the spleen in 2 ml of RPMI medium and filter cells through a 40 μm mesh into a 50 ml conical tube.
3. Centrifuge cells at 270 × *g* for 10 min and lyse red blood cells by incubating cells in red blood cell lysis buffer (5 ml) for 5 min (see Note 2).
4. Add 20 ml of PBS to stop lysis.
5. Centrifuge cells at 270 × *g* for 10 min and aliquot 1×10^8 cells in 5 ml PBS for fluorescence labeling.
6. Prepare a working dye stock of CellVue Maroon by adding 20 μl of dye in 5 ml of diluent C (provided by the company) and add to 5 ml cell suspension.
7. Incubate for 5 min at room temperature.
8. Add 10 ml RPMI/1% FBS to stop the reaction.
9. Centrifuge cells at 270 × *g* for 10 min.
10. Wash cell pellet three times with 10 ml RPMI/1% FBS.
11. Centrifuge cells at 270 × *g* for 10 min and resuspend cells in sterile PBS at a concentration of 2×10^6 cells/200 μl/mouse.
12. Splenocytes isolated from an uninfected mouse are injected intraperitoneal into an *H. pylori* infected host mice using a 25 gauge needle and 1 ml syringe.
13. Mice are anesthetized using isoflurane, and imaged on the In Vivo Multispectral Imaging System (Carestream Molecular Imaging, CMI) with exposure time 10 s (excitation filter 630 nm, emission filter 700 nm) and subjected to X-ray (35 KVP, 30 s exposure, 0.5 mm filter) (Fig. 1) (see Note 3). Mice are imaged every day up to 7 days post-adoptive transfer.

3.2. Beta-Galactosidase Detection

Genetic labeling of cells using the *Escherichia coli* beta-galactosidase gene (LacZ) is an accepted method used for the identification, localization, and engraftment of transplanted cells in vivo.

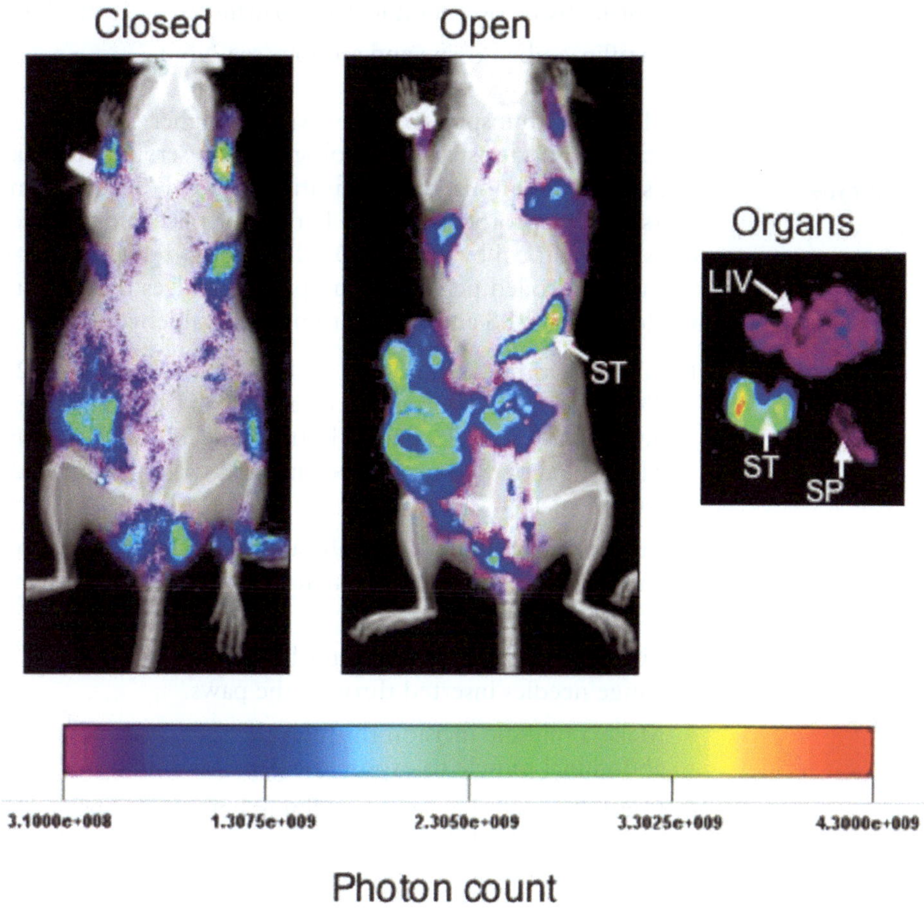

Fig. 1. Fluorescence in vivo imaging of labeled splenocytes 7 days post-adoptively transferred into *H. pylori*-infected mouse. The images shown are of a mouse closed (*left image*), opened (*middle image*) and dissected organs (*right image*) showing expression of labeled splenocytes within the stomach at the site of infection (also see Note 3). Mice may be imaged over several days by shaving the abdomen. The body cavity can be opened for anatomical localization of fluorescence (Fig. 1). *ST* stomach, *LIV* liver, *SP* spleen.

Stable expression of beta-galactosidase expression within cells may be detected using a sensitive immunohistochemical system for detecting LacZ. The protocol for beta-galactosidase staining was modified from a published method by Hendrikx et al. (9).

1. Collect stomach tissue and wash well in 1× PBS.
2. Pin tissue to a piece of dental wax placed in a well of a six well plate (see Note 4). Fix stomach tissue with 4% paraformaldehyde for 1 h on ice.
3. Wash tissue using rinse buffer for 30 min three times at room temperature.
4. Stain tissue from 4 h to overnight with X-gal staining solution at 37°C in the dark (see Note 5).
5. Wash tissue with rinse buffer for 30 min at room temperature.

6. Post-fix tissue overnight in 10% formalin at 4°C (see Note 6).
7. Paraffin-embed tissue and section (see Note 7).

3.3. Alternative Protocol for Fixing Tissue: Paraformaldehyde Perfusion

For detecting transplanted cells with a higher sensitivity while maintaining good gastric morphology and preserving other organs, perfusion of the mouse with a fixative for in vivo fixation may be necessary. The protocol detailed here is for mice and is based on a published protocol by Zeller (10). The method does not require a pressure-controlled perfusion pump and involves the exchange of the blood with PBS and then with paraformaldehyde (10).

1. Fill one syringe with 20 cm^3 1× PBS, and another with 20 cm^3 4% paraformaldehyde and store at 4°C on ice. The syringes will be attached to a 25 gauge butterfly needle with a stopcock. This will allow syringes to be changed without the need for a second stick.
2. Euthanize mouse by CO_2 inhalation under approved protocol and immediately lay animal on its back. DO NOT euthanize by cervical dislocation.
3. Pin the mouse to a Styrofoam board—using push pins or 18 gauge needles inserted through the paws.
4. Make a midline incision through the skin, and the underlying abdominal muscle to expose the peritoneal lining.
5. Carefully open the peritoneal lining being cautious not to nick the bowel. Pin the skin, muscle, and the lining to a Styrofoam board. This will immobilize the mouse and allow an easily viewed field.
6. Carefully cut through the rib cage and remove the diaphragm to access the heart. It is advised to work quickly but carefully. If the blood clots or the main blood vessels are damaged the perfusion cannot be performed.
7. Cut the right atria to allow blood to flow out.
8. Carefully insert the needle attached to the syringe filled with 1× PBS, into the left ventricle.
9. Slowly but constantly perfuse the 1× PBS into the heart (Fig. 2). You will know if the perfusion is working if the liver and spleen turn grayish-white.
10. After the blood has been flushed out, remove the syringe with 1× PBS and attach syringe filled with ice cold 4% paraformaldehyde. It is often helpful to have a second person help you with this step. Slowly perfuse the mouse with 20 ml paraformaldehyde. A successful perfusion is indicated by a muscle tremor that is observed on the limbs and tail. By the end of this procedure the animal should be stiff.
11. Following the perfusion, dissect out stomach and rinse out contents with PBS.

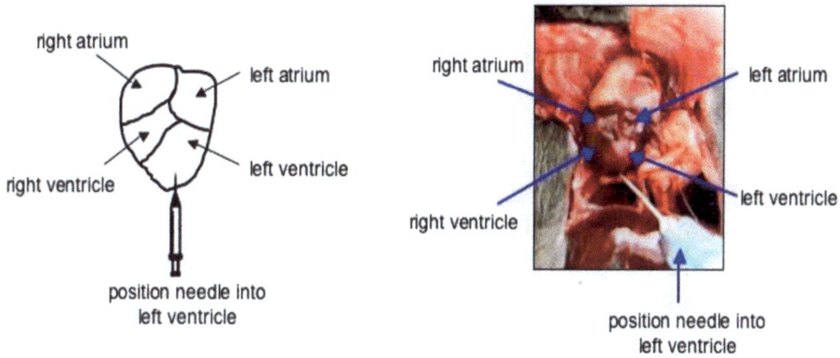

Fig. 2. Paraformaldehyde perfusion of the mouse. The diagram and photo illustrates the chambers of the heart and the positioning of the syringe in the left ventricle during perfusion.

12. Proceed with steps 3–7 outlined above for the beta-galactosidase detection.
13. If direct fluorescent imaging is indicated, the tissue is processed in OTC for frozen section. Alternately, tissue can be processed for routine histology.

3.4. X/Y FISH

The following method was modified using the package insert for the CEP X Spectrum Orange/Y Spectrum Green DNA Probe Kit as a guide (Abbot Molecular Inc., Product Number 30-161050/32-161050).

3.4.1. X/Y Paint Production

A variety of commercially available paints are available for X/Y-FISH. The Vysis CEP X/Y probe set (Abbott Molecular, Inc.) targets centromeric alpha-satellite Xp11.1-q11.1 and satellite III Yq12 loci yielding orange and green fluorescence, respectively. If commercial chromosome paints are not available for the sample being evaluated, then the following steps should be used to generate the necessary paints:
Isolate chromosomal material from the tissue sample or cell population of interest (using standard method for the chosen cell type). An example of such a purification may be found in Rens et al. (11).

1. Prepare the chromosomes for sorting by staining with 40 μg/ml of Chromomycin A3 (Sigma), 2 mM $MgSO_4$, and 2 μg/ml Hoechst 33258 for at least 2 h.
2. Immediately prior to sorting, add 10 mM sodium sulfite and 25 mM sodium citrate.
3. Collect X and Y chromosomal material in separate PCR tubes containing 30 μl of sterile water.
4. Use the chromosomes as a template for degenerate oligonucleotide primer (DOP) PCR.

5. Perform DOP-PCR using labeled nucleotides in order to generate "painted" product DNA. Biotin-16-dUTP and Cy3-dUTP are recommended.
6. Dilute the X and Y paints to 50 ng per 15 μl of hybridization buffer.

3.4.2. Interphase/ Metaphase Cell Preparation

Isolation of interphase or metaphase cell populations from sex-mismatched, bone marrow transplanted animals should be performed according to the AGT Cytogenetics Laboratory Manual (12).

1. Collect bone marrow samples in sodium-heparinized vacutainers and use immediately to prepare slides. Hemolyzed samples should not be used for FISH analysis.
2. Prewarm denaturing solution to 73°C.
3. Wash twice in iced PBS.
4. Resuspend 10^5 cells in 0.2 ml of ice cold PBS. (Note: A minimum of 100 μl is required for cytospin slide preparation, with the overage included if a concentration adjustment is needed to achieve the desired number of cells per image field. This volume should be scaled based on the number of replicates required.)
5. Prepare a cytospin device by placing slides and filters into the appropriate slots. The cardboard filter should face the center of the cytospin. Prewet each well with 100 μl of cold PBS and spin for 1–2 min.
6. Quickly add 100 μl (approximately 5,000 cells) of each sample to the appropriate wells of the cytospin and spin for 5 min at 500 rpm. Carefully remove the filters and slides being careful not to disrupt the pelleted cells.
7. Visually examine each slide using low power phase contrast microscopy (10× magnification) in order to confirm that the cells have pelleted properly and that each field contains at least 100 interphase cells. Air-dry slides.
8. Fix slides in Carnoy's fixative for 15 min and air dry.
9. Hold the prepared slides in a dark slide box at −20°C until needed for hybridization.

3.5. Fluorescence In Situ Hybridization Assay

3.5.1. Denaturation of Specimen DNA

1. Denature the specimen DNA by submerging the slides in 73°C denaturing solution for 5 min.
2. Remove slides from bath.
3. Dehydrate slides through a 1 min transfer series of 70, 95, and 100% ethanol baths.
4. Prepare all probe hybridization solutions and allow the X,Y paint probes to warm to room temperature.

5. Immediately before hybridization, drain the excess alcohol by wicking solution from the edge of the slide onto a paper towel.
6. Dry the slides on a 50°C slide warmer for 2 min and proceed immediately to the next step.

3.5.2. Hybridization

1. Vortex probe solution to mix. Denature mixture in a 75°C water bath for 5 min.
2. Obtain the desired test slides and appropriate low proportion male (5% XY and 95% XX) and low level female (5% XX and 95% XY) control smears.
3. Add 10 μl of the probe solution and 10 μl of hybridization buffer to each cytospin and coverslip. Air bubbles trapped under the coverslip interfere with hybridization and should be avoided.
4. Transfer slide(s) to a prewarmed, covered chamber and transfer to the hybridization oven. Hybridization may be performed for a minimum of 30 min at 42°C or, if preferred, overnight at 37°C (see Note 8).
5. Following hybridization, slides may be sealed (around the edge of the coverslip) with rubber cement and held in a humidified hybridization chamber overnight (<16 h) until final washes are performed.
6. Carefully remove the coverslips and wash for 5 min in 50°C 1× SSC buffer (see Note 9).
7. Transfer slides to 0.5× SSC for 5 min.
8. Quickly rinse slides in distilled water, wick off excess liquid, and air dry in the dark.
9. Coverslip and seal the slides using 20 μl of DAPI II mountant.
10. Store hybridized slides desiccated at –20°C for up to 1 year.

3.5.3. Analysis: Quality Verification

1. Scan slides under wide-field fluorescence using a 25× objective to confirm uniform, bright signal intensity, dark background intensity, and specificity of hybridization. Acceptable staining can present as either compact, brightly stained points or more dimly fluorescent, diffuse oval-shaped points. Elevated background or reduced signal-to-noise intensities may result from insufficient washing, hybridization stringency, or improperly high paint concentration. Additionally, probes may cross-hybridize to nontarget sequences if stringency is too low. If either of these situations occur, optimization of these parameters should be performed with positive and negative control slides used for each set of assay conditions.
2. A majority (>95%) of cells should display one or both of the *fluorescent signals.*

3.5.4. Analysis: Control Slide Requirements

1. View each control slide through a 25× or 63× objective. Image cells using fluorescence filter sets appropriate for the chosen X and Y paints or according to manufacturer instructions. (Imaging each emission channel sequentially will help to identify areas of signal overlap between fluorophores.)
2. Choose an area of the cell smear that is uniformly distributed with minimal cell overlap and clumping.
3. Count the number of signals from a total of 500 interphase nuclei and 20 metaphase spreads beginning in the upper left quadrant image field and scanning from left to right, top to bottom. Only *distinct* signals should be counted from interphase nuclei, and only nonoverlapping chromosomes with clear staining should be counted for metaphase spreads.
4. The assay should be repeated if less than 95% of the nuclei show specific staining.
5. Tally the number and percentage of interphase nuclei with XX and XY signals. Greater than 95% of the cells should be clearly identified as either XX or XY. The percentage of XX and XY cells should fall within the data sheets provided with the control slides.

3.5.5. Analysis: Sample Interpretation

1. Enumerate test slides as described above for the control slides.
2. Specimens with greater than 0.6% of the interphase cells identified as the donor gender are considered positive for the presence of donor cells (e.g., 3 out of 500 total cells). Similarly, if any of the 20–30 metaphase spreads are identified as having donor origin, the specimen is considered positive.

4. Notes

1. When this protocol is used with cancer cells the fluorescence of the dye starts to diminish 30 days post-transplant.
2. With incubation in red blood cell lysis buffer the solution should appear clear when red blood cells are lysed.
3. Mice may be imaged over several days by shaving the abdomen and for the final imaging, after euthanasia, body cavity can be opened for anatomical localization of fluorescence (Fig. 1.).
4. Stomach sections may be pinned onto a piece of dental wax to keep tissues flat during the fixing and staining procedure.
5. The tissue color may be monitored over this period of time. For strong signals tissues will become blue within 4 h of staining with X-gal solution. Staining may take up to 24 h. In our experience, any staining that appears after 24 h is nonspecific

and should be interpreted cautiously. Always interpret in the context of the appropriate negative control. Tissue collected from nontransplanted mice may be used as a negative control.

6. After fixing overnight in 10% formalin, tissue may be rinsed in PBS and stored in 70% ethanol at 4°C until paraffin-embedded.
7. After sectioning, tissue may be counterstained by either immunohistochemistry or immunofluorescence.
8. Hybridization may be performed for a minimum of 30 min; however, to increase the signal intensity, hybridization time may be increased to overnight.
9. For an alternative approach using less stringent 0.4× SSC, wash at elevated 73°C temperature.

Acknowledgments

The author would like to acknowledge the expert technical assistance of Kathleen LaSance (Lab Manager) and Dr. Lisa Lemen (Director) in the Vontz Core Imaging Laboratory (VCIL, University of Cincinnati). This work was supported by NIH 1R01DK083402-01A2 grant (Y. Zavros).

References

1. Houghton J, Li H, Fan X, Liu Y, Liu JH, Rao VP, Poutahidis T, Taylor CL, Jackson EA, Hewes C, Lyle S, Cerny A, Bowen G, Cerny J, Moore N, Kurt-Jones EA, Erdman SE (2010) Mutations in bone marrow-derived stromal stem cells unmask latent malignancy. Stem Cells Dev 19:115–1166
2. Houghton J, Nomura AB, Rogers SCS, Carlson J, Li H, Cai X, Fox JG, Goldenring JR, Wang TC (2004) Gastric cancer originating from bone marrow-derived cells. Science 306:1568–1571
3. Quante M, Tu SP, Tomita H, Gonda T, Wang SS, Takashi S, Baik GH, Shibata W, Diprete B, Betz KS, Friedman R, Varro A, Tycko B, Wang TC(2011) Bone marrow-derived myofibroblasts contribute to the mesenchymal stem cell niche and promote tumor growth. Cancer Cell 19:257–272
4. Takaishi S, Okumura T, Wang TC (2008) Gastric cancer stem cells. J Clin Oncol 26:2876–2882
5. Mor G, Yin G, Chefetz I, Yang Y, Alvero A (2011) Ovarian cancer stem cells and inflammation. Cancer Biol Ther 11(8):708–713
6. Fang D, Nguyen TK, Leishear K, Finko R, Kulp AN, Hotz S, Van Belle PA, Xu X, Elder DE, Herlyn M (2005) A tumorigenic subpopulation with stem cell properties in melanomas. Cancer Res 65:9328–9337
7. Singh SK, Hawkins C, Clarke ID, Squire JA, Bayani J, Hide T, Henkelman RM, Cusimano MD, Dirks PB (2004) Identification of human brain tumour initiating cells. Nature 432:396–401
8. Takaishi S, Okumura T, Tu S, Wang SS, Shibata W, Vigneshwaran R, Gordon SA, Shimada Y, Wang TC (2009) Identification of gastric cancer stem cells using the cell surface marker CD44. Stem Cells 27:1006–1020
9. Hendrikx PJ, Vermeulen J, Hagenbeek A, Vermey M, Martens AC (1996) LacZ staining in paraffin-embedded tissue sections. J Histochem Cytochem 44:1323–1329

10. Zeller R (2001) Fixation, embedding, and sectioning of tissues, embryos, and single cells. Curr Protoc Mol Biol, Chapter 14, Unit 14.1
11. Rens W, Yang F, Welch G, Revell S, O'Brien PC, Solanky N, Johnson LA, Ferguson Smith MA (2001) An X-Y paint set and sperm FISH protocol that can be used for validation of cattle sperm separation procedures. Reproductuion 121:541–546
12. Barch MJ et al (1997) The agt cytogenetics laboratory manual, 3rd edn. Lippincott-Raven, New York

Chapter 26

In Vivo Measurement of *Helicobacter pylori* Infection

Marjan Mohammadi, Samaneh Saberi Kashani, Yeganeh Talebkhan Garoosi, and Sahar Jahangiri Tazehkand

Abstract

Helicobacter pylori is a well-recognized gastroduodenal pathogen (National Institute of Health Consensus Conference, JAMA 272:65–9, 1994) and a class I carcinogen (International Agency for Research on Cancer, IARC Monograph on the Evaluation of Carcinogenic Risk to Humans 61:177–240, 1994) which successfully colonizes the harsh acidic environment of the stomach. *H. pylori* is the causative factor for peptic ulcer disease (PUD) and an independent risk factor for gastric adenocarcinoma development. Therefore, accurate detection of infection is crucial for devising proper eradication regimens and preventing the more severe GI complications.

Detection of *H. pylori* in the gastric mucosa can be performed via (1) direct detection of the bacterium; culture, histology and polymerase chain reaction (PCR) or (2) indirect detection of its enzymatic products particularly urease as well as serum *H. pylori*-specific antibody responses, which can be detected by rapid urease test (RUT) and serology, respectively.

The accuracy of these diagnostic tests is reported as follows: 98.1% for bacterial culture, 98.1% for histology, 94.3% for PCR, 96.2% for RUT, and 84.9% for serology (Ni et al, J Pediatr 136(6):823–7, 2000).

Key words: *Helicobacter pylori*, Transfer medium, Culture medium, RUT, Serology, Stool-PCR

1. Introduction

Bacterial culture is usually restricted to research settings, or unusual clinical situations. The culture of *H. pylori* is a preferable method for the following: (1) Diagnostics: to determine *H. pylori* antibiotic sensitivity and devising proper eradication regimens; (2) Research: to investigate microbial host-cell interaction, pathophysiology, growth requirements, and metabolism; and (3) Production: for development of diagnostic, preventive, and therapeutic products (1, 2).

H. pylori culture from gastric biopsy specimens is not popular among clinical laboratories because it is time-consuming, costly, and

JeanMarie Houghton (ed.), *Helicobacter Species: Methods and Protocols*, Methods in Molecular Biology, vol. 921, DOI 10.1007/978-1-62703-005-2_26, © Springer Science+Business Media, LLC 2012

requires skilled laboratory personnel adept at culturing this demanding and fastidious organism. Culture disadvantages include requirements for (1) special transport conditions, (2) rapid sample processing, (3) costly and complicated culture media with specific incubation conditions, (4) lengthy time before confirmation and devising treatment regimens for patients.

H. pylori is a fastidious microorganism which requires a microaerophilic environment and complex nutrient media for its growth. Several selective and differential media have been evaluated and recommended (3–8). These media are usually based on Columbia, Brucella, or brain heart infusion agar base containing blood or blood products with an additive such as starch, charcoal, cyclodextrin, or bovine serum albumin (BSA) (9). These media, depending on the conditions, can be supplemented with antimicrobial agents to prevent the overgrowth of contaminants, present in the gastrointestinal tract and/or the endoscopy/transportation environment.

Rapid Urease Test (RUT) is a reliable indirect means of detecting viable *H. pylori* organisms in the gastric lumen, and is performed on gastric biopsy specimens obtained at the time of upper endoscopy. This test is preferred by endoscopists due to its ease of use, rapidity and cost-effectiveness. The principle behind this test is that *H. pylori* produces abundant amounts of cell surface-associated urease as a catalytically active heterodimer. In the gastric environment urease catalyzes urea into carbon dioxide and ammonia. The produced hydroxyl ions then react with carbon dioxide, producing bicarbonate, which neutralizes gastric acid providing a means for successful colonization of *H. pylori* in the otherwise harsh acidic gastric environment (10). The RUT takes advantage of the urease of the organism, and the fact that *H. pylori* is the only urease producing organism found in/on the gastric mucosa. At the time of diagnostic upper endoscopy, several biopsy specimens are taken from the antrum and the fundus of the stomach. The specimens are placed into a prepared and commercially available medium containing urea and an indicator such as phenol red. The urease produced by *H. pylori* hydrolyzes urea to ammonia, which raises the pH of the medium. This pH change is indicated by a change in the color of the specimen from yellow (NEGATIVE) to red (POSITIVE). Other pH indicators can be used, and the time to development of a color change ranges from 1 to 24 h, and is dependent upon the particular kit used.

Serology is the most widely used method of detecting *H. pylori* infection and most applicable for noninvasive population screening approaches, and sometimes (though less useful) for monitoring treatment success. (11). Disadvantages of serology include positive results due to the prolonged persistence of *H. pylori* specific antibodies despite its successful eradication (12) or cross reactivity of antibody to other pathogens giving an apparent positive result, which however does not reflect an active or past infection with *H. pylori* (13).

Because of these findings, previous studies have recommended several *H. pylori* antigens/immunogens for incorporation into serology assays. These antigens are presented as components of intact cells, sonicated cell extracts as well as purified antigens in native or recombinant forms (14). Extensive antigenic heterogeneity among *H. pylori* strains isolated from different geographic locations has created demand for development of homemade assay incorporating antigens from local clinical strains. Among the various *H. pylori*-specific serology assays, homemade enzyme linked immunosorbent assay (ELISA) tests using native whole cell extracts have demonstrated the highest diagnostic accuracy among various serologic assays (15).

Stool-based PCR can detect *H. pylori* DNA in stool specimens (16) in a noninvasive manner most suitable for screening children and elderly. In comparison to other reliable and noninvasive methods which are expensive and technically demanding, stool-based PCR seems efficient and cost effective. However, its efficacy is influenced by fecal PCR inhibitors such as complex polysaccharides which inhibit the further processing of purified DNA (17). To remove inhibitors a variety of DNA purification methods have been developed; these involve use of filters (18), organic extraction and ethanol precipitation (19) and modified commercial kits which have met with varying success. Here, we describe a noninvasive and nonproprietary stool PCR protocol based on previously developed methods (19–21).

Comparison of the available tests for detection of *H. pylori* show that in the absence of eradication therapy, serology is the method of choice because of its cost efficiency and ease of performance making it applicable in low-tech laboratories as a powerful tool in epidemiological screening programs. If a reinfection is suspected, or if successful eradication needs to be investigated, systemic tests such as urea breath tests (UBT) and stool antigen assays are preferable over serology because they can accurately detect current infection.

2. Materials

2.1. Bacterial Culture and RUT

2.1.1. Rut Medium

1. 0.1 g yeast extract, 20 g urea, 0.091 g monopotassium phosphate (KH_2PO_4), 0.095 g disodium phosphate (Na_2HPO_4), and 0.01 g phenol red. Dissolved in 800 ml dH_2O, bring volume up to 1 l and adjust to pH 6.9.
2. Sterilize the solution by passing it through a 0.45 μm filter.
3. Aliquot solution in sterile microtubes (600 μl/tube) and store stock solution in dark bottles at 4°C for a maximum of 48 h. However, it is best used fresh (see Notes 1–3).

2.1.2. Transfer Medium

1. Solution A: Dissolve 2 g casamino acid, 2 g peptone, 0.4 g yeast extract, 1 g NaCl, and 0.32 g agar in 150 ml dH_2O and

bring up to a total volume of 200 ml. Autoclave at 120°C, 15 psi (1.05 kg/cm^2) for 20 min, cool and store at 4°C.

2. Solution B: Dissolve 0.04 g L-Cysteine in 20 ml dH_2O and adjust pH to 7.0. Add 0.2 g glucose to the solution and bring up the volume to a total of 40 ml, sterilize the solution by passing it through 0.45 μm filter (Millipore) and store at 4°C.
3. Combine solutions A and B and add 28 ml autoclaved glycerol (87%) to the combination.
4. Aliquot the transfer medium in sterile microtubes (600 μl/tube). This medium can be stored at 4°C for up to 6 months (see Note 4).

2.1.3. HPSPA Culture Medium (H. pylori Special Peptone Agar) (see Note 5)

1. Rehydrate specified amounts of Brucella agar (as directed by the manufacturer) in 920 ml dH_2O (for a final volume of 1 l).
2. Add 3 g yeast extract and 5 g beef extract as growth supplements.
3. Autoclave at 120°C, 15 psi (1.05 kg/cm^2) for 20 min. Cool down to 56°C in a water bath (see Note 6).
4. Dissolve 0.5 g ferrous sulfate in 4 ml dH_2O, sterilize by passing through 0.22 μm filters and add to the cooled culture medium.
5. Dissolve 0.5 g sodium pyruvate in 4 ml dH_2O, sterilize by passing through 0.22 μm filters and add to the cooled culture medium. (see Note 7)
6. Prepare the following stock solutions in dH_2O and sterilize by passing through 0.22 μm filters:
   (a) 5 mg/ml fungizone.
   (b) 5 mg/ml vancomycin.
   (c) 10 mg/ml trimethoprim.
7. Add to the culture media 400 μl of fungizone, 1.2 ml of vancomycin, and 500 μl of trimethoprim from these stock solutions to create the following final concentrations: 6 mg/l vancomycin, 20 mg/l trimethoprim, and 2 mg/l fungizone.
8. Finally add 70 ml (7%) defibrinated sheep blood. Mix gently and dispense 15–20 ml per plate (15 × 100 mm) under sterile conditions.
9. Following solidification at room temperature, stock and store in plastic bags at 4°C for up to 1 week (see Notes 8 and 9).

2.2. Helicobacter pylori Identity Tests

2.2.1. Crystal Violet Staining Solution

1. Dissolve 2 g crystal violet in 20 ml 95% ethanol.
2. Dissolve 0.8 g ammonium oxalate in 80 ml dH_2O.
3. Mix the above two solutions. The mixture is stable for 2–3 years at room temperature.

2.2.2. Iodide Staining Solution

1. Dissolve 2 g potassium iodide in 2–3 ml dH_2O (the crystals will dissolve and the solution temperature will fall).
2. Dissolve 1 g iodine crystals in the concentrated potassium iodide solution. Dilute the mixture to 300 ml with dH_2O.

2.2.3. Destaining Solution (Acetone–Alcohol)

1. Mix 100 ml acetone (reagent grade) with 100 ml 95% ethanol. Store in a brown-glass bottle, label with 1 year expiration date, and store at room temperature.

2.2.4. Counterstaining Solution

1. Dissolve 0.25 g safranin in 10 ml 95% ethanol. Dilute the mixture to 100 ml with dH_2O.

2.2.5. Catalase Test (See Note 10)

1. Prepare 3–6% hydrogen peroxide solution (supplied in various concentrations by commercial manufacturers) (see Notes 11 and 12).

2.2.6. Oxidase Test

1. There are different discs that are commercially available and may contain *N,N,N′,N′*-tetramethyl-*p*-phenylenediamine (TMPD) or *N,N*-Dimethyl-*p*-phenylenediamine (DMPD) (BBL, Difco, REMEL) which turn dark blue to maroon when oxidized.

2.2.7. Wet Mount

1. Clean microscopic slide.
2. dH_2O.
3. Flame-sterilized loop.
4. Light microscope.

2.3. ELISA Reagents and Buffers

Dissolve 8 g NaCl, 0.2 g KCl, 1.1 g Na_2HPO_4, and 0.2 g KH_2PO_4 in 800 ml H_2O.

2.3.1. Phosphate Buffered Saline

1. Adjust the pH to 7.2 (using HCl). Bring up the volume to 1 l with dH_2O.
2. Autoclave the solution at 121°C for 20 min at 15 psi (1.05 kg/cm^2) and store at 4°C.

2.3.2. Protease Inhibitor Solution

1. Dissolve one tablet (protease inhibitor cocktail, Roche, Germany) in 10 ml phosphate buffered saline (PBS) (see Note 13).

2.3.3. Coating Buffer (Carbonate–Bicarbonate Buffer)

Dissolve 1.59 g Na_2CO_3, 2.93 g $NaHCO_3$, and 0.01% (w/v) sodium azide (NaN_3) in 800 ml dH_2O (see Note 14). Adjust the pH to 9.6 with NaOH or HCl. Bring up the volume to 1 l with dH_2O and store at 4°C. The coating buffer can be stored at 4°C for up to 2 weeks.

2.3.4. Blocking Buffer

Dissolve 1% (w/v) bovine serum albumin in PBS (see Note 15). Freshly prepared buffer is highly recommended.

2.3.5. Diluent Buffer

1. Mix 0.1% (v/v) Tween 20 in desired amounts of blocking buffer (see Note 16). Freshly prepared buffer is highly recommended.

2.3.6. Standards

1. Standard A: *H. pylori* negative serum.
2. Standard B (cutoff control): 10 U/ml anti-*H. pylori* specific IgG (IBL, Germany).
3. Standard C (weakly positive control): 25 U/ml anti-*H. pylori* specific IgG (IBL, Germany).
4. Standard D (strongly positive control): 150 U/ml anti-*H. pylori* specific IgG (IBL, Germany).

2.3.7. Washing Buffer

1. Mix 0.1% (v/v) Tween 20 in PBS. Washing buffer can be stored at 4°C for up to 2 weeks.

2.4. Stool-Based PCR

2.4.1. Sample Collection

1. Devices: Any clean and dry container such as a bedpan, plastic plate or bag, newspaper, and a capped container.

2.4.2. DNA Analysis

1. Phosphate-buffered saline: Mix 0.01 M phosphate buffer, 0.0027 M potassium chloride and 0.317M sodium chloride. Adjust pH to 7.4.
2. Phenol–chloroform–isoamyl alcohol (PCI): (25:24:1).
3. Sodium acetate and ethanol.
4. Ultrapure water.
5. RNase ONE: Promega, Southampton, UK.
6. TE buffer: Tris-acetate–EDTA buffer (10 mM Tris–HCl, 1 mM EDTA). Adjust pH to 7.4.

2.4.3. Gene Capture

1. Guanidine thiocyanate.
2. Capture probe: *H. pylori*-specific biotinylated capture probe (HpS1 (5-GGG GAG TAC GGT CGC AAG ATT AAA ACT CAA AGGAAT A-3)), which targeted the 16S rRNA gene of *H. pylori* (20).
3. Paramagnetic polystyrene beads coated with Streptavidin: Dynabeads M-280–Streptavidin.
4. Wash buffer: 0.1 M Tris–HCl, 0.01 M EDTA, 1 M sodium chloride, and 0.1% (v/v) Tween 20.

2.4.4. PCR

1. Reaction buffer: (Tris–HCl 10 mM, KCl 50 mM).
2. $MgCl_2$ (1.5 mM).
3. Deoxynucleoside triphosphates (200 μM).
4. Forward and reverse primers: HpF (5-GCG ACC TGC TGG AAC ATT AC-3) and HpR (5-CGT TAG CTG CAT TAC TGG AGA-3).
5. Hot Start DNA polymerase.

3. Methods

3.1. Bacterial Culture and RUT

The entire handling of this section should be conducted under a horizontal laminar-flow cabinet with a high-efficiency particulate air (HEPA) filter which provides an almost sterile working environment. Flame sterilization can also be used to further decontaminate the immediate working environment, in order to (1) eliminate any externally introduced contaminants from the exposed sides of media bottles, culture flasks, or test tubes and (2) sterilize small instruments such as forceps, wire inoculating loops, etc.

Gastric specimens are collected during gastroscopy; preferably according to Sydney protocols (22) (see Note 17).

3.1.1. RUT Analysis

Place the biopsy obtained from the Incisura Angularis in microtubes containing RUT medium by sterile forceps (see Note 18). The color change of RUT medium from pale yellow to deep pink is considered as a positive test result (see Notes 19 and 20).

3.1.2. Culture of Gastric Specimens

1. Place gastric biopsies in microtubes containing transfer media by sterile forceps and transfer them to the bacterial culture lab within 4 hours (see Note 21).
2. Mince and homogenize biopsies with tissue grinder in the transfer medium prior to culture.
3. Dispense and spread 300 μl of the thoroughly homogenized biopsies on HPSPA culture plates (see Note 22).
4. Incubate at 37°C under microaerobic conditions (see Note 23).
5. Turn over culture plates after 1 day when homogenized biopsies are absorbed to the culture media (see Note 24).
6. Check media for detection of *H. pylori* colonies after 5–7 days of incubation (see Notes 25 and 26).

3.2. Helicobacter pylori Identity Tests

3.2.1. Gram Staining

1. Smear preparation (see Note 27): Place a drop of sterile saline or water on a microscopic slide, transfer a sample of *H. pylori* colony using a sterile applicator. Gently mix to emulsify. Avoid mixing vigorously and creating aerosols.
2. Air dry smears on a flat surface. Fix the slides by passing them two or three times through a flame. To avoid distortions, do not overheat. Allow the slides to cool before staining.
3. Immerse the fixed slides in the crystal violet solution; allow the stain to remain for 30 s.
4. Discard the solution, and rinse slide gently under running water (see Note 28).
5. Rinse off excess water with iodine solution, and then flood the slide with fresh iodine solution. Allow the iodine to remain for 30 s.

6. Rinse gently under running water.
7. Decolorize by letting the destaining solution flow over the smear while the slide is held at an angle. Stop when the runoff becomes clear. Remove excess decolorizer with gentle flow of running water (see Note 29).
8. Flood the slide with the counterstaining solution and allow it to remain for 30 s.
9. Remove excess counterstain with a gentle flow of running water.
10. Drain slides and air dry in an upright position. Clean off the bottom of the slide by wiping on a paper towel.
11. Examine the smear under microscope which will reveal *H. pylori* microorganisms as gram-negative curved rods.

3.2.2. Catalase Test

1. Place approximately 0.2 ml of hydrogen peroxide solution in a test tube.
2. Carefully pick an *H. pylori* colony with an applicator (see Note 30) and place above the surface of the hydrogen peroxide solution.
3. Cap the tube and tilt to allow the hydrogen peroxide solution to cover the colony.
4. Look for vigorous bubbling to occur within 10 s which confirms the presence of catalase-positive *H. pylori*.

3.2.3. Oxidase Test

1. Moisten oxidase discs with sterile dH_2O and place them in a Perti-dish and add a loop full of *H. pylori* colony from the culture plate.
2. Presence of oxidase producing *H. pylori* will sequentially change the color of disks to pink, through maroon and into black, within 10–30 s.

3.2.4. Wet Mount

Wet mount is a test for confirmation of *H. pylori* structure and motility.

1. Place a drop of dH_2O in the middle of a microscopic slide.
2. Allow the flame-sterilized loop to cool before transferring a small portion of a single colony to the drop and dissolving it.
3. Cover with a slip and view under light microscope.
4. *H. pylori* microorganisms will appear as spiral-shaped bodies with cork-screw spiraling motility.

3.3. ELISA

The following steps should be performed on ice:

3.3.1. Antigen Solution

1. Harvest *H. pylori* colonies from the surface of 5–10 (3–5 day old) culture plates with 1 ml PBS per plate into corresponding number of microtubes (see Note 31).

2. Centrifuge the bacterial suspension at $2,300 \times g$ at 4°C for 10 min. Remove the supernatants and resuspend the pellets in equivalent volumes of PBS (1 ml) and repeat the centrifugation procedure.
3. Combine and measure the wet weight of the pellet and resuspend it in sufficient volume of protease inhibitor solution to make the final concentration of 0.5 g/ml.
4. Sonicate the bacterial suspension at 20,000 Hz on ice five times for 45 s with 45 s intervals.
5. Centrifuge the bacterial sonicate at $7,200 \times g$ for 1 h at 4°C. Carefully transfer the supernatant to a new tube and discard the pellet (see Note 32).
6. Sterilize the obtained supernatant by sequentially passing it through 0.8, 0.45 and 0.22 μm (Millipore) filters. This step will also remove bacterial cell debris.
7. Measure and label the protein concentration of antigen solution at 280 nm by spectrophotometer (a protein concentration of 30–50 mg/ml is expected).
8. Aliquot and store at −20°C until further use.

All buffers and reagents should be brought to room temperature (21–25°C) prior to use (see Note 33).

3.3.2. Antigen Coating

1. Select the number of needed wells/strips according to Fig. 1.
2. Place the assigned number of strips into an ELISA microwell frame (Maxisorb ELISA plate, Nunc, Denmark).
3. Dilute antigen solution a final concentration of 5 μg/ml with coating buffer. Add 100 μl to each microwell. Cover the frame with a plastic seal and incubate it at 4°C over night.
4. Wash the coated wells three times with PBS (300 μl/well, see Note 34). Invert the plate and gently tap it on a clean dry paper towel after each washing step.

3.3.3. Blocking

1. Add 100 μl blocking buffer to each previously coated well (see Note 35), cover the frame with a plastic seal and incubate for 1 h at room temperature.
2. Remove the blocking solution and tap the inverted plate on a clean, dry paper towel.

3.3.4. Serum and Standard Addition

1. Dilute test sera at a 1:100 rate using the diluent buffer (e.g., 2 μl + 198 μl).
2. Add 100 μl of standards (A–D) and diluted test sera to each assigned well (Fig. 1). Cover the frame and incubate at room temperature for 1 h.

	1	2	3	4	5
A	Standard A	Patient#5			
B	Standard B	Patient#6			
C	Standard C	Patient#7			
D	Standard D	Patient#8			
E	Patient#1	Patient#9			
F	Patient#2	etc.			
G	Patient#3	etc.			
H	Patient#4	etc.			

Fig. 1. Map of ELISA plate.

3. Remove serum samples/standards and wash the wells five times with the washing buffer (300 μl/well). After the final wash, tap the plate on clean, dry paper towel to remove all liquids from the wells.

3.3.5. Conjugate Antibody Addition

1. Dilute polyclonal rabbit anti-human IgG conjugated with Horseradish Peroxidase (HRP, Dako, Denmark) at a 1:10,000 rate in the diluent buffer.
2. Add 100 μl diluted conjugated antibody to each well, cover the frame and incubate at room temperature for 1 h.
3. Remove the conjugated antibody and wash the wells five times with the washing buffer (300 μl/well). After the final wash, tap the plate on clean, dry paper towel to remove all liquids from the wells.

3.3.6. Substrate Addition

1. Add 100 μl of ready to use Tetramethylbenzidine (TMB, Sigma) solution to each well.
2. Cover the frame and incubate at room temperature for 10 min.
3. Avoid direct exposure to light during incubation (see Note 36).

3.3.7. Stopping the Reaction

1. Add 100 μl stop solution (2 M H_2SO_4) to each well in the same order as substrate addition (see Notes 37 and 38).

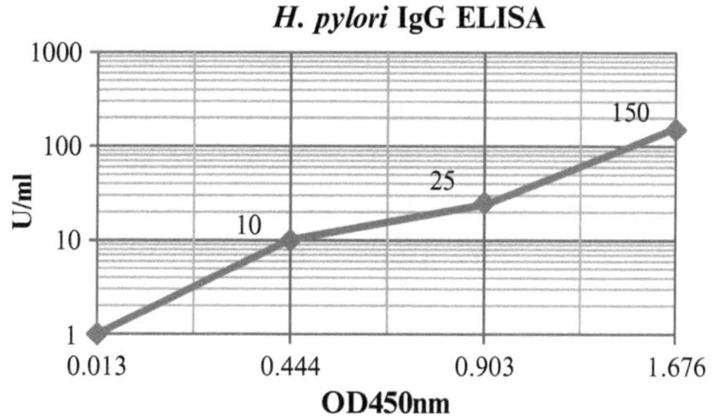

Fig. 2. Standard curve.

3.3.8. Reading

1. Read the strips at 450 nm by a microtiter plate reader capable of reading absorbance at 450 nm.
2. Dispose the plate after reading.

3.3.9. Interpretation of the Results

1. The cut off value is determined by the optical density (OD) measured for Standard B. Samples in the range of 20% above or below the cut off value are considered as borderline (see Note 39). Optical densities higher and lower than the borderline zone are interpreted as positive and negative readings, respectively.
2. For quantitative determination of antibody titers, calculate the estimated IgG concentration of each serum sample according to standard readings. Graph the OD values of standards on the *X* axis (linear) and their IgG international units on the *Y* axis (logarithmic) on a semi-logarithmic graph paper. Based on the obtained OD value of each serum sample, its concentration of anti-*H. pylori* IgG can be deduced from the standard curve (Fig. 2).

3.4. Stool-Based PCR

3.4.1. Sample Collection

1. Using a collection device, collect the fecal sample, transfer to a capped container, and store at −80°CC until analysis.

3.4.2. DNA Analysis

1. Homogenize fecal samples immediately after melting the specimen in sterile phosphate-buffered saline in a Stomacher 400 (Seward Medical, London, UK) for 10 min at room temperature to produce 20% (w/v) fecal solution.
2. Centrifuge the solution at 20,000 × *g* for 30 min. The pellet should remain undisturbed.
3. Transfer the supernatant to a sterile 2-ml tube and add an equal volume of a sample preparation reagent, PrepMan Ultra (Applied Biosystems, Cheshire, UK) (see Notes 40–43).

4. Boil samples for 10 min; cool for 2 min, and then centrifuge at 20,000 × *g* for 10 min. (see Note 44).
5. Add an equal volume of phenol–chloroform–isoamyl alcohol (PCI) (25:24:1) to supernatant and mix by inversion then centrifuge at 20,000 × *g* for 3 min (see Note 45).
6. Transfer the upper phase to another 2-ml tube, and repeat the process.
7. Add an equal volume of chloroform, mix by inversion then centrifuge at 20,000 × *g* for 3 min (see Note 46).
8. Collect total nucleic acid by precipitation with 0.1 volume of cold (0°C) 3 M sodium acetate and two volumes of cold (0°C) 100% ethanol on ice for 30 min and then centrifuge at 16,000 × *g* for 15 min.
9. Dry in air, and resuspend samples in 266 μl of ultrapure water.
10. Add 40 U of RNase ONE and incubate for 1 h at 37°C to remove RNA (see Note 47).
11. Precipitate DNA with sodium acetate and ethanol as described before (see Note 48).
12. Resuspend the harvested DNA in TE buffer.

3.4.3. Gene Capture

1. Combine total fecal DNA suspensions with 300 μl of 6 M guanidine thiocyanate and 20 nM of the capture probe and incubate overnight at 25°C (see Note 49).
2. Harvest *H. pylori* DNA by using 10 μl of paramagnetic polystyrene beads coated with Streptavidin and wash three times in wash buffer (see Note 50).
3. Incubate 6 min at 85°C for final harvesting, and then transfer *H. pylori* DNA to a clean tube.

3.4.4. PCR

1. Prepare PCR master mix as follows: Reaction buffer (Tris–HCl 10 mM, KCl 50 mM), $MgCl_2$ (1.5 mM), deoxynucleoside triphosphates (200 μM), forward and reverse primers at 0.5 μM each, 1.25 U of Hot Start DNA polymerase, and 10 μl of target DNA (final reaction volume, 50 μl). The following primers are suggested: HpF (5-GCG ACC TGC TGG AAC ATT AC-3) and HpR (5-CGT TAG CTG CAT TAC TGG AGA-3) (see Note 51).
2. Using the following program amplify template DNA: An initial denaturation at 94°C for 5 min (see Note 52); 40 cycles of denaturation at 94°C for 1 min, annealing at 60°C for 1 min, and extension at 72°C for 1 min; and a final extension at 72°C for 5 min.

3. Resolve PCR products (138 bp) by agarose gel electrophoresis (3% agarose in 0.5× Tris-acetate–EDTA buffer) for 1 h at 50 V.
4. Stain the gel with ethidium bromide.
5. Examine DNA bands under UV illumination.

4. Notes

1. Do not heat urea base into solution as it decomposes upon heating (23).
2. Urea is known to undergo autohydrolysis; therefore, it is advisable to store urea-based media at 4–8°C. Color change may take slightly longer when media is refrigerated (23).
3. RUT media exposed to light may develop peroxide, which could interfere with the urease reaction (23).
4. Prior to use, make sure the media is clear with no turbidity. You can also check for contamination by incubation of an aliquot at 37°C overnight.
5. *H. pylori* can grow on different media containing blood supplements. Most studies have used Brucella or Columbia as the agar base. An amount of 7 to 10% blood cause better growth of *H. pylori* as compared with 5% blood. Horse blood better supports *H. pylori* growth as compared to sheep blood (8).
6. 56°C is the optimum temperature for addition of blood supplements before solidification.
7. Do not heat or autoclave sodium pyruvate or ferrous sulfate.
8. For contamination control, place a sample plate at 37°C and check for growth. Lack of growth authorizes the use of remaining plates for *H. pylori* culture (24).
9. The added antibiotics will select for the growth of *H. pylori* from gastric specimens but will also limits the rate of growth. For the subsequent bacterial passages, the addition of antibiotics is optional (25).
10. *H. pylori* produce catalase which is detected by decomposition of hydrogen peroxide into oxygen and water ($2H_2O_2 \rightarrow 2H_2O + O_2$) (26).
11. Hydrogen peroxide is unstable and should be stored at 4°C, avoiding light.
12. Hydrogen peroxide can cause irritation and is harmful if swallowed.
13. Protease inhibitor cocktail is used for the inhibition of several metalloproteases in cellular extracts of bacteria, mammalian,

yeast and plant cells including serine and cysteine metalloproteases. One tablet is sufficient for the inhibition of the proteolytic activity in 10 ml protein extraction solution.

14. Sodium azide (NaN_3) is used for increasing the shelf life of material by preventing bacterial contamination. It is a severely irritating material. Solutions containing NaN_3 must be clearly labeled. Disposal of solutions containing NaN_3 should be followed by large volumes of water to avoid interaction with the plumbing system. Handling should be performed with great care using safety gloves and goggles. Upon exposure, thoroughly wash the area with water. NaN_3 can also inhibit the activity of conjugated antibodies; therefore, use fresh tips for addition of antibodies.
15. Avoid harsh agitation of blocking buffer to prevent BSA proteolysis.
16. Tween 20 is a non ionic detergent which binds to water insoluble components rendering them hydrophilic.
17. The majority of gastric sampling is obtained from the antral region. However, for optimal results obtaining several biopsies from various gastric locations is preferred (22).
18. Incisura angularis is the sharp angular depression in the lesser curvature of the stomach which separates the fundus from the antrum (22).
19. An increase in pH due to the production of ammonia results in a color change from yellow (pH 6.8) to bright pink (pH 8.2) (23).
20. It is recommended that RUT tubes be monitored for the desired color change for up to 24 h. Bacterial culture is however recommended despite RUT negative results (27).
21. Lengthy transportation periods reduces culture yield, particularly if the patient has previously received antibiotic treatment or the numbers of colonizing *H. pylori* in the gastric biopsies are minimal, in which case may result in no bacterial growth or false negative results. This situation can be remedied by prolonged culture incubation for up to 12 days.
22. Spreading tissue homogenate on the entire plate surface increases the rate of *H. pylori* recovery and prevents the overgrowth of potential contaminants (24).
23. In general, primary cultures of *H. pylori* are sensitive to oxygen in comparison with most Campylobacter species. *H. pylori* grows in microaerobic atmosphere (2–5% O_2, 5–10% CO_2 and 0–10% H_2) in CO_2 incubators or microaerobic chambers with gas-generation kits (28).

24. Watch for contaminations by daily monitoring of culture plates. In case contaminants appear, subculture of *H. pylori* colonies onto fresh plates is recommended.
25. The question of how long cultures should be incubated has been evaluated by several investigators, and regardless of the type of media or the selected atmosphere, plates should be kept for up to 10 days before they can be considered negative for primary isolation of *H. pylori*. The length of incubation also depends on whether selective or nonselective media is used (24).
26. *H. pylori* growth appears as yellowish 0.5 to 2 mm colonies on horse blood agar and 0.5 to 1 mm pale grayish colonies on sheep blood agar.
27. A suitable bacterial smear is a thin monolayer of bacterial culture on slides which will enables visualization of bacterial morphology under microscopic examination.
28. Excessive rinsing during this step could cause detachment of crystal violet from gram-positive cells. To ensure a gentle flow across the smeared side, apply the flow of water to the underside of the angled slide.
29. A properly decolorized smear appears with an almost olive-green tone and without observable evidence of crystal violet.
30. Catalase test should not be performed on colonies mixed with media containing red blood cells which may cause false positive results.
31. Bacterial colonies can be previously collected during several bacterial passages, in order to yield sufficient bacterial proteins. To avoid bacterial lysis the collected strains should be kept at –70°C.
32. It is important to avoid collecting pellet material into the supernatant as it will block the sterilizing filters.
33. Use latex gloves throughout the procedure. Avoid cross contamination of reagents as it may produce false results.
34. Insufficient washing will result in assay variations affecting the validity of test. Completely remove the wash buffer after the last wash (remove bubbles).
35. Avoid bubbles upon addition of buffers as they may yield false results.
36. Presence of *H. pylori*-specific antibodies will turn the colorless substrate solution into blue.
37. Addition of the stop solution will transform the blue color to yellow. The stopped reaction can be read up to 1 h.
38. Handle sulfuric acid under chemical fume hood as this acid is severely irritating to the skin, eyes and the respiratory tract. It may create severe damages including fires upon contact with other material. Upon exposure, thoroughly wash the area with water.

39. Samples with borderline values are reported as equivocal or borderline after having repeated the test.
40. PrepMan Ultra has been formulated for the removal of PCR-quality DNA from foodstuffs.
41. PrepMan Ultra is a chemical hazard; it contains material that may irritate eye, skin, and respiratory tract, and cause adverse effects on the kidneys, blood and central nervous system. Wear appropriate protective eyewear, clothing, and gloves.
42. In this step Gramley et al. (19) used an organic extraction method to remove inhibitors from stool by which they mixed the sample with 8 M urea containing 1% sodium dodecyl sulfate, 20 mM Tris–HCl (pH 8.0), 100 mg of Chelex (Bio-Rad, Hercules, Calif.), and 50 mg of polyvinylpyrrolidone followed by incubation at 60°C.
43. An alternative method is to homogenize 4 g of frozen feces in 28 ml of EXACT Buffer A (Exact Sciences Corporation), then centrifuge twice (20). Another method is to dissolve 1 g of stool sample in 100% ethanol and chloroform and wash with acetone after centrifugation (19).
44. As an alternative method, Gramley et al. (19) mixed supernatant with 0.7 M sodium chloride and 1% hexadecyl trimethyl ammonium bromide (CTAB) (Sigma) and incubated at 65°C after a step of boiling and alcohol precipitation.
45. This step will separate total nucleic acid from the protein content.
46. This treatment will remove the remained PCI in the liquid phase
47. Gramley et al. (19) performed organic extraction and alcohol precipitation prior to incubation with RNase A (1 mg/ml; Sigma) and proteinase K (0.5 mg/ml; Bio-Rad) at 58°C for 2 h.
48. In the final step, Gramley et al. (19) performed another round of organic extraction and alcohol precipitation with a solution of 3 mM Tris–HCl (pH 7.5) and 0.2 mM EDTA.
49. According to Shuber et al (20), it is also possible to mix the extracted DNA with an equal volume of 6 M guanidinium isothiocyanate and 20 nM of capture probe.
50. Ahlquist et al (29) performed an additional hour of incubation at room temperature in this step.
51. All experiments must include both a positive control (such as purified *H. pylori* DNA) and a negative control (such as substitution of the template DNA with sterile water).
52. In case using Hot start DNA polymerase enzyme a prolonged heating (15 min) at this temperature is required to activate the enzyme.

References

1. Mégraud F (1997) Diagnosis and candidates for treatment of Helicobacter pylori infection. How should Helicobacter infection be diagnosed? Gastroenterology 113:S93–S98
2. Lee A (1998) Animal models for the host-pathogen interaction studies. Br Med Bull 54:163–173
3. Queiroz DMM, Mendes EN et al (1998) Indicator medium for C. pylori. J Clin Microbiol 25:2378–2379
4. Dent JC, McNulty CAM (1988) Evaluation of a new se lective medium for C. pylori. Eur J Clin Microbiol Infect Dis 7:555–568
5. Glupczynski Y, Labbe M, Thibaumont F (1989) Comparative Evaluation of a New Selective Culture Medium for Improved Isolation of *Campylobacter pylori* from Gastric Biopsy Specimens. In: Megraud F, Lamouliatte H (eds) Gastroduodenal pathology and *Campylobacter pylori*. Elsevier, New York, pp 3–6
6. Westblom TU, Madan E et al (1991) Egg yolk emulsion agar, a new medium for the cultivation of H. pylori. J Clin Microbiol 29:819–821
7. Cellini L, Allocati N et al (1992) New plate medium for growth and detection of urease activity of H. pylori. J Clin Microbiol 30:1351–1353
8. Olivieri R, Bugnoli M et al (1993) Growth of H. pylori in media containing cyclodextrins. J Clin Microbiol 31:160–162
9. Goodwin CS, Worsley BW (1993) Microbiology of Helicobacter pylori. Gastroenterol Clin North Am 22:5–19
10. Hu LT, Mobley HL (1990) Purification and N-terminal analysis of urease from Helicobacter pylori. Infect Immun 58:992–998
11. Xie F, Luo N, Lee HP (2008) Cost effectiveness analysis of population based serology screening and (13) C-Urea breath test for Helicobacter pylori to prevent gastric cancer: a Markov model. World J Gastroenterol 14:3021–3027
12. Hirschl AM, Rotter ML (1996) Serological tests for monitoring Helicobacter pylori eradication treatment. J Gastroenterol 31(suppl IX):33–36
13. Malfertheiner P, Megraud F, O'Morain C et al (2007) Current concepts in the management of Helicobacter pylori infection-The Maastricht III Consensus Report. Gut 56:772–781
14. Herbring P, van Doorn LJ (2000) Serological methods for diagnosis of Helicobacter pylori infection and monitoring of eradication therapy. Eur J Clin Microbiol Infect Dis 19:164–173
15. Leal YA, Flores LL, Garcia-Cortes LB et al (2008) Antibody-based detection tests for the diagnosis of Helicobacter pylori infection in children: a meta-analysis. PLoS One 3:e3751
16. Mapstone NP, Lynch DA et al (1993) PCR identification of H. pylori in faeces from gastritis patients. Lancet 341:447
17. Monteiro L, Bonnemaiso D et al (1997) Complex polysaccharides as PCR inhibitors in feces: Helicobacter pylori model. J Clin Microbiol 35:995–998
18. Cavallini A, Notarnicola P et al (2000) Use of macroporous polypropylene filter to allow identification of bacteria by PCR in human fecal samples. J Microbiol Methods 39:265–270
19. Gramley WA, Asghar A et al (1999) Detection of Helicobacter pylori DNA in fecal samples from infected individuals. J Clin Microbiol 37:2236–2240
20. Shuber AP, Ascano JJ et al (2002) Accurate, noninvasive detection of Helicobacter pylori DNA from stool samples: potential usefulness for monitoring treatment. J Clin Microbiol 40(1):262–264
21. MacKay WG, Williams CL (2003) Evaluation of protocol using gene capture and PCR for detection of Helicobacter pylori DNA in feces. J Clin Microbiol 41(10):4589–4593
22. Dixon MF, Genta RM (1996) Classification and grading of gastritis. The updated Sydney system. International workshop on the histopathology of gastritis, Houston 1994. Am J Surg Pathol 20(10):1161–1181
23. Macfaddin JF (2000) Biochemichal tests for identification of medical bacteria. Urease test, 3rd edn. Lippincott Williams and wilkins, Philadelphia, PA
24. Perez-Perez GI (2000) Accurate diagnosis of Helicobacter pylori culture including transport, Gastroenterol Clin North Am 4:879–884
25. Miendje Deyi VY, Van den Borre C (2010) Comparative evaluation of 3 selective media for primary isolation of Helicobacter pylori from gastric biopsies under routine conditions. Diagn Microbiol Infect Dis 68(4): 474–476
26. Manos J, Kolesnikow T (1998) An investigation of the molecular basis of the spontaneous occurrence of a catalase-negative

phenotype in Helicobacter pylori. Helicobacter 3(1):28–38
27. Versalovic J (2003) Helicobacter pylori pathology and diagnostic strategies. Am J Clin Pathol 119:403–412
28. Andersen LP, Kiilerick S et al (1998) An analysis of seven different methods to diagnose Helicobacter pylori infections. Scand J Gastroenterol 33:24–30
29. Ahlquist DA, Skoletsky JE et al (2000) Colorectal cancer screening by detection of altered human DNA in stool: feasibility of a multitarget assay panel. Gastroenterology 119:1219–1227

Index

JeanMarie Houghton (ed.), *Helicobacter Species: Methods and Protocols*, Methods in Molecular Biology, vol. 921, DOI 10.1007/978-1-62703-005-2, © Springer Science+Business Media, LLC 2012

MIX
Papier aus verantwortungsvollen Quellen
Paper from responsible sources
FSC® C105338

If you have any concerns about our products,
you can contact us on
ProductSafety@springernature.com

In case Publisher is established outside the EU,
the EU authorized representative is:
Springer Nature Customer Service Center GmbH
Europaplatz 3, 69115 Heidelberg, Germany

Printed by Libri Plureos GmbH
in Hamburg, Germany